# THE CAMBRIDGE REGION

EDITED BY

H. C. DARBY

*Ehrman Fellow of King's College, Cambridge,*
*and Lecturer in Geography in the University*

CAMBRIDGE
AT THE UNIVERSITY PRESS
1938

CAMBRIDGE UNIVERSITY PRESS
Cambridge, New York, Melbourne, Madrid, Cape Town,
Singapore, São Paulo, Delhi, Mexico City

Cambridge University Press
The Edinburgh Building, Cambridge CB2 8RU, UK

Published in the United States of America by Cambridge University Press, New York

www.cambridge.org
Information on this title: www.cambridge.org/9781107687479



First published 1938
First paperback edition 2013

*A catalogue record for this publication is available from the British Library*

ISBN 978-1-107-68747-9 Paperback

# THE CAMBRIDGE REGION

This is an extream pleasant open Country, and a place of such Variety and Plenty, that fruitful *Ceres* with a smiling Countenance, invites the Industrious Peasant to behold with Joy the Fruits of his Labour, whilst she crowns his Industry with a plentiful Harvest; and as if the Earth strove not to be behind-hand with him in conferring other Largesses, she in divers places makes some Annual Additions of another Crop, by adorning the Fields with large Productions of Saffron, by which great Profits do continually arise: Besides, here it is that the green Banks of murmuring Rivers, and sunny Hills bedeck'd with diversity of Plants and Simples, call forth the Students from their musing Cells, and teach them Theory as well as Practice, by diving into their Natures, contemplating their Signatures, and considering their Qualities and various Effects. In a word, here is nothing wanting for Profit or Delight; and though the Northern parts of the County towards the Isle of *Ely*, lying somewhat low, are moist and Fenny, yet that Defect is abundantly supply'd by the Plenty of Cattle, Fish, and Fowl bred in those Fenns, and which makes the Air more healthy, the gentle Gales which are frequently stirring, drive away all thick Mists and Fogs which in some parts most annoy it, and by this means it is become a fit Seat for the Muses to inhabit, and we have no reason to complain of the Soil, since our Wise Ancestors thought it good and convenient to plant a Colony of Learned Men here, and place one of the Eyes of our Nation in this spot of Ground, the famous and most glorious University of *Cambridge*, which we could not in Honour pass by without a Visit.

JAMES BROME, *Travels over England, Scotland and Wales* (1700), pp. 57–8.

# PREFACE

This survey of the district around Cambridge is written mainly in terms of the two administrative counties of Cambridge and the Isle of Ely. For, despite many disadvantages, a county is a convenient unit to consider—if only for statistical and historical reasons. The boundaries of these two units in Cambridgeshire include two types of country:

(1) The Upland, comprising both clay and chalk country. To this area some measure of cohesion is given by the valley of the Cam and its tributaries (see Fig. 3).

(2) The Fenland, marked by a contrast between the peat and silt areas, and including the fen islands (see Fig. 47).

Where necessary, we have not hesitated to extend discussion beyond the limits of the county boundaries. This is particularly true of the two chapters on the Fenland, for the problems of the Cambridgeshire fens cannot be discussed apart from wider considerations.

Finally, there is a third type of country—the Breckland—with a marked individuality; this comes within the county boundary only just west of the River Kennett. Chapter XV, however, is concerned with the Breckland as a whole—an area that lies almost entirely outside the county. Despite its artificial basis, this selection of material has the advantage of including the three types of country within easy access of the town of Cambridge itself (see Fig. 56).

The survey was prepared, in the first instance, for the Cambridge Meeting (1938) of the British Association for the Advancement of Science, and was presented by the Syndics of the University Press to members attending the Meeting. Its preparation was made possible by the kindness of thirty-four contributors, some of whom are indebted to the Editor of the *Victoria County Histories* (Mr L. F. Salzman) for the use of material assembled for the forthcoming volumes on Cambridgeshire; that indebtedness is specifically acknowledged on the relevant pages below. Most of the maps were drawn by Mr L. D. Lambert; others by Mr D. Baldwin and Mr E. A. Ashman. Permissions to reproduce the material on the maps are acknowledged on p. xiii. The courtesy of many people in the county who have helped in various ways must also be remembered; likewise, the efficiency of the printers who cheerfully endured many things.

H. C. DARBY

KING'S COLLEGE
CAMBRIDGE
*Lammas Day* 1938

## CONTENTS

# MAPS AND DIAGRAMS

# ACKNOWLEDGMENTS

We are indebted for the following permissions:

For Figs. 1, 2, 3, 8, 23, 24, 25, 26, 34, 35, 37, 39, 49 and 52 to the Controller of H.M. Stationery Office and the Director General of the Ordnance Survey.

For Figs. 4, 29 and 38 and portions of Figs. 18, 19, 20, 21, 47, 48 and 56 to the Controller of H.M. Stationery Office and the Director of the Geological Survey.

For Figs. 54 and 55 (reproduced from British Admiralty Charts, nos. 1177 and 108) to the Controller of H.M. Stationery Office and the Hydrographer of the Navy.

For the information on Figs. 33 and 36 to the Ministry of Agriculture and Fisheries.

For Fig. 7 to Major Gordon Fowler and the Editor of the *Geographical Journal.*

For Fig. 16 to the Editor of the *Journal of Ecology.*

For Figs. 17 and 22 to the Editor of the *Proceedings of the Cambridge Antiquarian Society.*

For access to the information on Figs. 30 and 31 to the Editor of the *Victoria County Histories.*

For Fig. 32 to Mr J. H. Wardley.

For Fig. 40 to Dr J. A. Venn.

For Fig. 46 to Mr W. P. Spalding.

For Figs. 50 and 51 to the Chief Engineer of the River Great Ouse Catchment Board.

For Fig. 57 to the Land Utilisation Survey.

For Fig 58 to the Editor of *Economic History.*

H. C. D.

CHAPTER ONE

# THE GEOLOGY AND PHYSIOGRAPHY OF THE CAMBRIDGE DISTRICT

Edited by O. T. Jones, F.R.S.

(With contributions by W. G. V. Balchin, A. G. Brighton, E. C. Bullard, H. Godwin, O. T. Jones, W. V. Lewis, and T. T. Paterson)

IN CAMBRIDGESHIRE AND THE SURROUNDING COUNTRY (FIG. 1), three broadly contrasted areas can be readily distinguished: (i) the Chalk escarpment to the east and south-east; (ii) the western plateau; and (iii) the Fenland occupying the northern part of the district. The area is drained mainly by the Cam and the Ouse, which flow from the upland, through the Fenland to the outfall at King's Lynn. In the Fens they are joined from the east by the Lark, the Little Ouse, and the Waveney, which drain the Chalk region east of Mildenhall, Brandon, and Stoke Ferry.

(i) *The Chalk Escarpment*, in the south and east, reaches its greatest height (549 ft. above O.D.) near Therfield, south-west of Royston; it declines north-eastward (to 400 ft. and below) towards Bury St Edmunds, and descends to still lower levels farther north-east. This watershed is crossed by three main depressions. One of these is followed by the Cambridge-Liverpool Street branch of the L.N.E.R. from Chesterford to Newport; the other by the Cambridge-King's Cross branch between Hitchin and Stevenage; while the third lies some miles to the east, and joins the valley of the Little Ouse with that of the Waveney.

The escarpment is determined largely by the Chalk Rock which outcrops near its brow; the overlying Upper Chalk leads down to Eocene beds on the fringe of the London Basin and is almost wholly covered by glacial drift.

From the low ground occupied by the Gault around Cambridge, the Chalk rises in gentle undulations to the brow of the escarpment (Fig. 4). Among these undulations the effect of certain hard bands in the Chalk, such as the Totternhoe Stone (or Burwell Rock) and the Melbourn Rock, can be distinguished by minor escarpments and dip slopes. The general character of this area is that of rounded ridges with intervening hollows carrying, at the present time, little surface drainage. It is traversed by shallow coombes (mainly dry valleys) which trend in a general north-west–south-east direction. The Gogmagog ridge (rising to 222 ft.) is a prominent feature near Cambridge (see Fig. 8).

Around Mildenhall, Brandon, Thetford and Lakenheath, the characteristic features of the Chalk upland are modified by a covering of gravels and sands that occupy the area known as Breckland. In places, deep circular depressions formed by solution reveal the presence of the underlying Chalk. Some of these depressions are very large and contain water more or less permanently, e.g. the Devil's Punch Bowl, 4 miles north of Thetford. Other depressions, which are probably also solution hollows, hold the well-known meres of the Thetford district, remarkable for the fluctuations of their water levels, and even for their complete desiccation at certain periods.

(ii) *The Western Plateau* lies south of Madingley along the Cambridge-Bedford road. From Eltisley, the plateau extends southwards for about 7 miles and occupies a considerable area; the surface stands between 200 and 250 ft., and, when viewed from a distance, looks remarkably even. Originally, it had a much wider extent both eastwards towards the lower part of the Chalk scarp, and westwards: its present limits are the result of dissection by the streams of the Ouse and Cam drainage systems. Large areas have, however, escaped dissection. The plateau is due in the main to a covering of Chalky Boulder Clay which overlies rocks ranging from Oxford Clay to Chalk. These rocks are exposed only on the dissected slopes of the plateau.

(iii) *The Fenland* occupies the northern part of the district. At one time the Chalk uplands of Norfolk were continuous with those of Lincolnshire, stretching across what is now the Wash. This ridge of Chalk was worn away by the action of rivers and the sea, and behind the ridge the Fenland was carved out of soft Jurassic clays—the Oxford, Ampthill and Kimeridge Clays. The surface of this Jurassic plain was uneven, and its higher portions projected above the general level to become the "islands" of the historical period. The whole region emerged from the various phases of the Ice Age with its islands capped with Boulder Clay, but with its basin nature unchanged. Subsequent time has witnessed the filling up of this basin.

The modern surface, therefore, is composed of post-glacial deposits consisting of alternating layers of peat and silt or clay (Buttery Clay), which rest on a foundation of formations ranging from Oxford Clay to glacial or post-glacial sands and gravels (March Gravels, etc.).

## RIVER SYSTEM

Broadly, the Cambridge district may be regarded as an immature peneplain, in which the influence of the varying resistance to erosion of the formations has not yet been obliterated. The area was invaded more than once by ice which has left behind a cover of Chalky Boulder Clay. The removal of this drift cover has re-exposed the rock formations and led to renewed differential erosion. During the retreat of the ice, deposits of

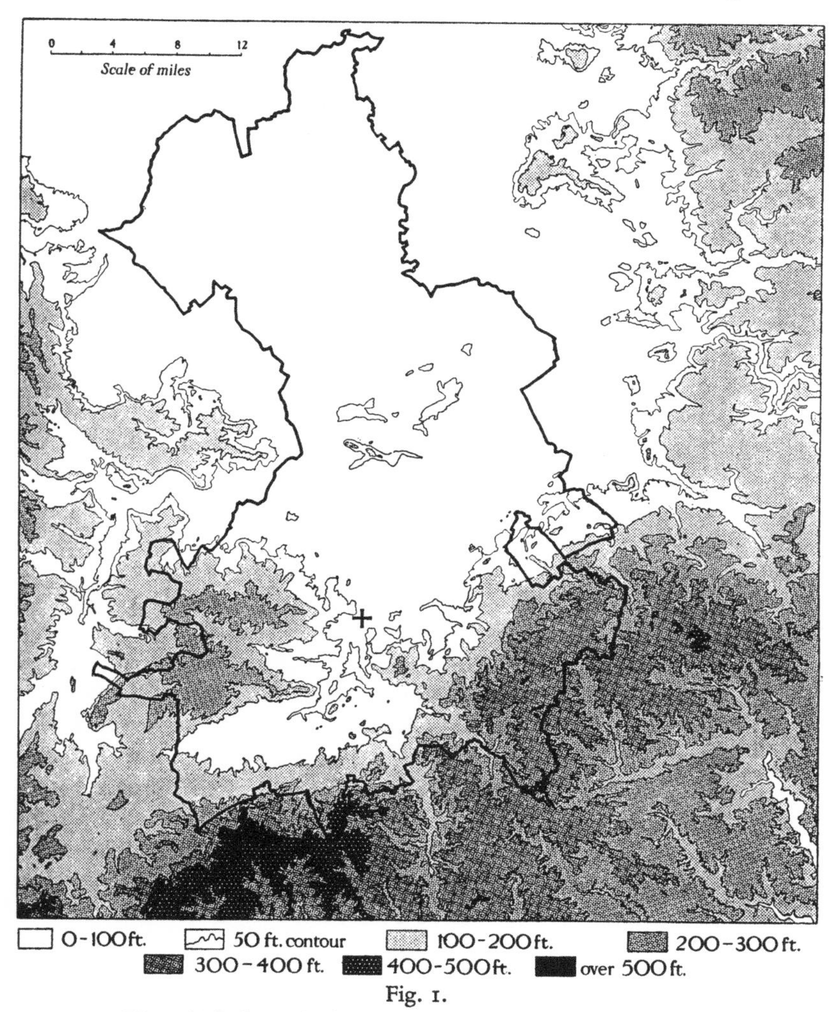


Fig. 1.

The relief of Cambridgeshire and the surrounding country.

coarse gravels and sands were laid down probably in glacial lakes. These now occur as ridges, which were formerly attributed to a system of river valleys older than the present Cam drainage. Gravels and sands near the eastern end of the Gogmagog ridge have been explained as outwash products from an ice sheet with an ice-contact slope along the south-west margin (Linton and Hildersham).

In the south of the area, the drainage consists of the tributaries of the River Cam which join to form the main river above Cambridge (Fig. 3). The Rhee, also known as Cam or Rhee, rises in the group of powerful springs which issue below the road at Ashwell (Herts), and flows as a strike stream mainly on the Gault to Trumpington, where it is joined by the Cam (also known as Cam or Granta), which rises about 4 miles south of Newport, and flows northward along the Chesterford-Newport depression.[1] This depression is underlain by a deep, narrow channel eroded into the Chalk to a depth of at least 20 ft., and possibly even to 100 ft., below O.D. It is filled largely by loams and sands, apparently laid down in water. A third tributary, known as the Granta, descends across the face of the Chalk escarpment from the neighbourhood of Bartlow and Linton. Of these, the Rhee and the Granta are related to the dip and strike of the rocks, whereas the Cam has been determined by the events that formed the above-named depression. The Bourn brook, which drains the western plateau, rises near Eltisley and enters the Cam above Cambridge, just below its junction with the Rhee (see Fig. 3).

It is not improbable that the three major depressions which traverse the Chalk escarpment were eroded by streams flowing from the north and west as consequent streams down the general dip slope of the Chalk, and that the headwaters of these streams were captured by the development of consequent or strike streams (e.g. the Rhee), leaving these depressions as wind gaps. The Chesterford-Newport deep channel was probably eroded further by an overflow from a glacial lake occupying the ground near Cambridge which was hemmed in between the ice on the north and the Chalk escarpment on the south. The Little Ouse-Waveney gap may also have been an overflow channel. In various parts of the district borings have revealed the existence of channels or holes eroded to considerable depths below O.D., but the origin of these is as yet unexplained. The valleys of the main rivers Cam and Ouse are occupied by river gravels in which certain well-defined terraces can be observed.

## UNDERGROUND STRUCTURE[2]

Although no rock older than part of the Great Oolite Series immediately beneath the Cornbrash outcrops within the area, older rocks are known from borings, particularly near Methwold where Middle and Lower Lias

[1] The Upper Cam has apparently brought about the capture of streams which formerly flowed southwards to the London Basin.

[2] By E. C. Bullard, Ph.D.

were proved; the latter to a depth of 660 ft. below O.D. Eastward, however, at Culford and near Harwich, Cretaceous rocks rest directly upon much older strata assumed to be Lower Palaeozoic; the Jurassic system has disappeared. It would be of considerable interest to know the depth and nature of this Palaeozoic floor. To this end, Bullard, Kerr-Grant, and Gaskell have applied the refraction seismic method with some success, and have obtained the results which are summarised briefly in Fig. 2.

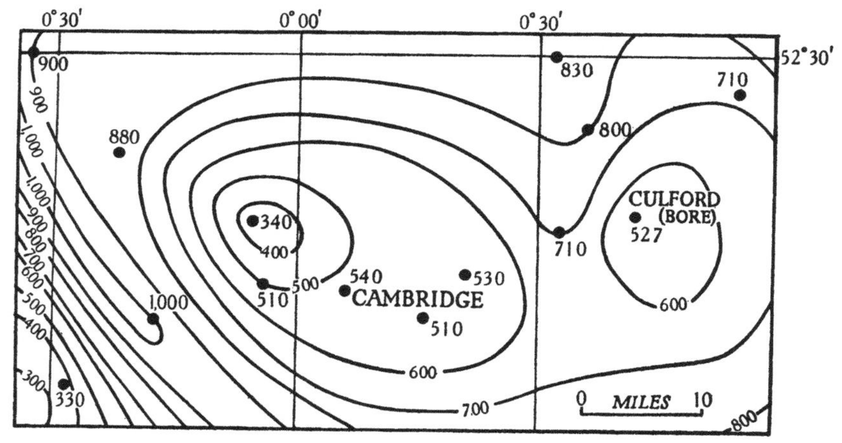


Fig. 2.

Depth of the Palaeozoic Floor in feet below O.D. Observations at further stations show that the valley to the west is less deep than is shown by the contouring above.

The stations shown above are as follows:

| Cambridge and to the west | | East of Cambridge | |
|---|---|---|---|
| Benefield | 900 | Bridgeham | 710 |
| Bourn (Cambs) | 510 | Culford (bore) | 527 |
| Cambridge | 540 | Feltwell | 830 |
| Fenstanton | 340 | Fulbourn | 510 |
| Houghton Conquest | 330 | Kentford | 710 |
| Leighton | 880 | Lakenheath | 800 |
| Tempsford | 1000 | Swaffham Prior | 530 |

(Figures in feet below O.D.)

These results are provisional and may be slightly different from the values finally published. The experimental error, due to inaccuracies of measurement and difficulties of interpretation, is of the order of 50–100 ft. The form of the contours round the margins of the map is based to some extent on a number of seismic stations and bores which lie outside the area.

## THE MESOZOIC ROCKS OF CAMBRIDGESHIRE[1]

The Mesozoic rocks of Cambridge strike roughly north-east and south-west, with a very gentle dip to the south-east.[2] The older beds (Jurassic) thus occupy the north and west of the County,[3] the younger beds (Cretaceous) the south-east. A few outliers of Cretaceous beds interrupt the Jurassic outcrops, as at Haddenham and Ely; and an anticlinal fold produces a Jurassic inlier surrounded by Cretaceous at Upware. The Cretaceous-Jurassic boundary is an unconformity; the base of the Cretaceous rests on the Kimeridge Clay at Ely, but, when traced to the south-west, this base oversteps the Kimeridge and Corallian in turn on to the Oxford Clay. These Jurassic rocks form the northern limb of an anticline with its axis (in the region of Sandy, Beds) pitching south-east.

The formations in the County may be summarised as follows:

| | | |
|---|---|---|
| Cretaceous | Chalk | |
| | Cambridge Greensand | |
| | Gault | |
| | Lower Greensand | |
| Jurassic | Kimeridge Clay | |
| | Corallian | Ampthill Clay, Coral Rag, Coralline Oolite |
| | | Elsworth Rock Series |
| | Oxford Clay | |

The Jurassic rocks are covered by drift in the Fens, except when they protrude to form the "islands" of Haddenham and Ely. The Cretaceous beds are largely covered by glacial deposits, both in the south-west and in the south-east (see Figs. 4 and 29).

### *OXFORD CLAY*

*Oxford Clay* is found in the west of the County and beyond, but it is badly exposed. A little beyond the County boundary at Forty Feet Bridge, north-east of Ramsey (Hunts), the Geological Survey have recently collected ammonites identified by Dr Spath as *Scarburgiceras scarburgense* (Young and Bird); but the best exposure is to the south-east of Ramsey at Warboys, where the same ammonite is plentiful. The Oxford Clay is dark blue or grey, with a few thin argillaceous limestones, bands of septarian nodules, selenite and pyrites (the fossils are often pyritised). Dr Arkell has suggested that the Warboys exposure is of the *mariae*-zone;

[1] By A. G. Brighton, M.A.

[2] To the north, in Norfolk, the strike changes to approximately north and south.

[3] I am indebted to the Director of the Geological Survey for permission to include some unpublished information about the Jurassic Clays in the Fenland. Part of the area is now undergoing revision by the Geological Survey.

lower zones of the Oxford Clay are, however, worked in the well-known brick pits around Peterborough.

The relation of the Oxford Clay to the Corallian requires further research. Dr Morley Davies has recognised a non-sequence just over the boundary at Sandy (Beds), where Corallian *Exogyra nana* beds rest on the *renggeri*-zone. The Elsworth Rock Series at Upware rest on the *mariae*-zone, as does the Corallian Limestone at Warboys.

### *CORALLIAN*

The correlation table given below is based on the work of Dr Arkell.

| Zone | Main outcrop | Upware inlier |
|---|---|---|
| *pseudocordata* | Upper Ampthill Clay (Long Stanton) | |
| "*variocostatus*" | Middle Ampthill Clay with Boxworth Rock | ? Coral Rag |
| *plicatilis* | Lower Ampthill Clay (Gamlingay) Upper Elsworth Rock Series | Coral Rag and Coralline Oolite |
| *cordatum* | Lower Elsworth Rock Series | Elsworth Rock Series |

The *Elsworth Rock* is a hard ferruginous calcareous mudstone, blue in colour when fresh, and brown when weathered. It is associated with clays and sandy beds, and all the rock-types contain limonite ooliths. It is diachronic. At Upware, it is 16 ft. thick, rests on Oxford Clay (*mariae*-zone), and belongs to the *cordatum*-zone, corresponding in age to the iron-shot clays which form the lower part of the series at Elsworth. The upper part of the series at Elsworth is about 12 ft. thick; from the abundance of ammonites, the masses of Serpulae and encrusting oysters, and the ironshot lithology, it is usually regarded as a condensed deposit. To the north, at Warboys, the base of the Corallian is represented by 3 ft. of hard shelly limestone with *Exogyra nana*, and a few, as yet undescribed ammonites; this is probably equivalent to part of the Elsworth Rock Series. To the south, at Sandy, the series is replaced by clays with limestone bands.

The *Ampthill Clay* is not well exposed. The pits at Gamlingay are overgrown; the excavation at Long Stanton, described by Dr Arkell, was temporary. New pits have been opened near Manea and Mepal (where the clay includes a limestone band, possibly the Boxworth Rock). Ampthill Clay fossils have been collected by the Geological Survey at Horseway on the Forty-Foot Drain, and at Honey Bridge on the Sixteen-Foot Drain. The Ampthill Clay is darker than the Oxford Clay, and contains phosphatic

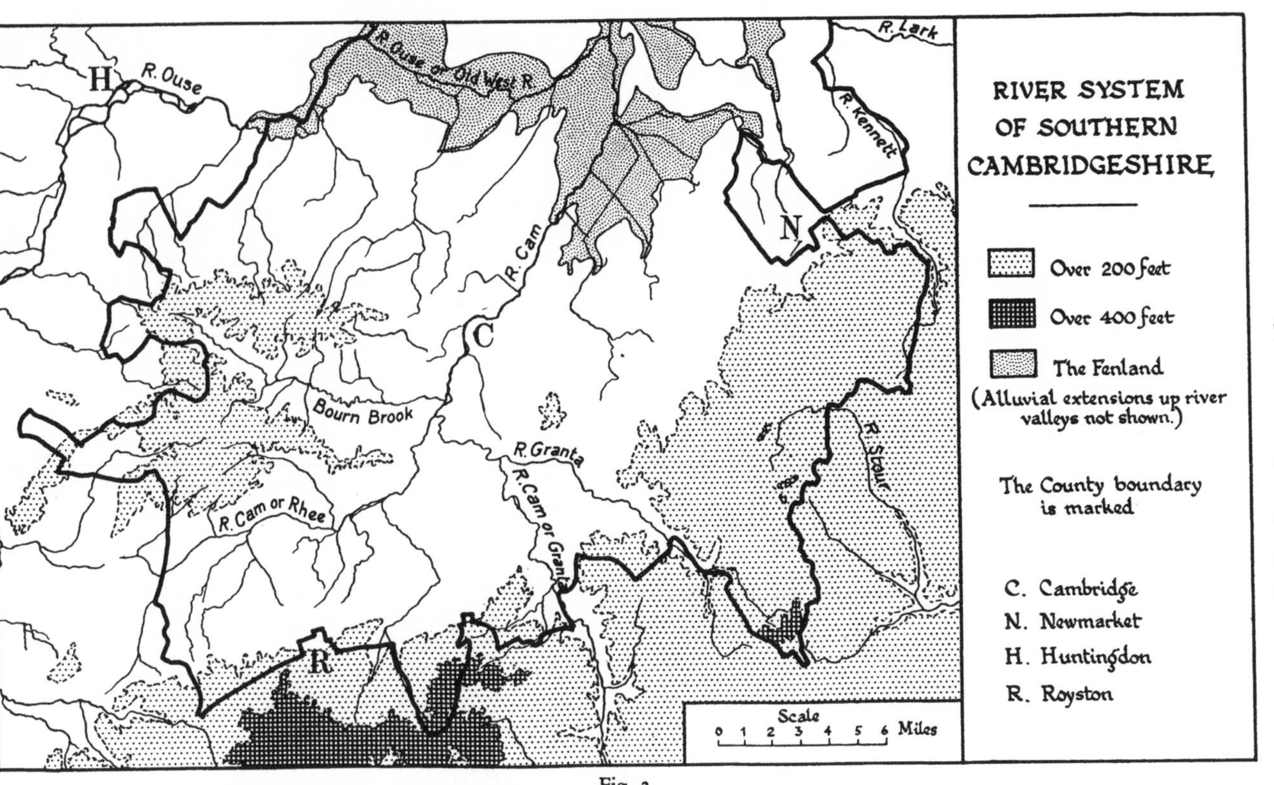


Fig. 3.
The River System of southern Cambridgeshire.

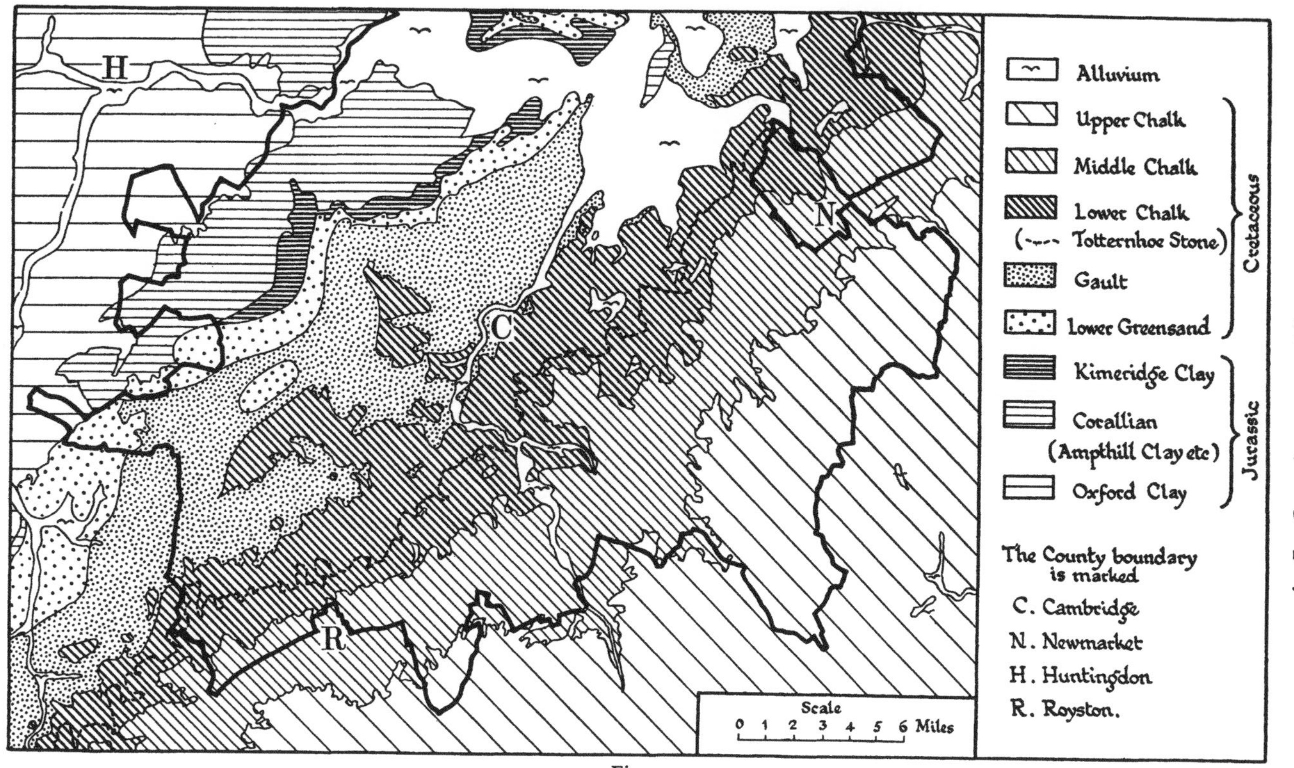


Fig. 4.
The solid Geology of southern Cambridgeshire (from information supplied by the Geological Survey).

nodules but little or no pyrites; limestone bands occur, some made up almost entirely of *Serpula intestinalis*. The fauna is muddy-water molluscan in character, with very rare echinoids in the limestone bands. To the east, in borings near Southery in West Norfolk, the Ampthill Clay is about 70–80 ft. thick.

*Coral Rag* and *Coralline Oolite* are known only at Upware, where Corallian limestone protrudes through the Cretaceous and extends over an area of about 3 miles by 1 mile. The Coralline Oolite is a cream-coloured limestone, full of small cavities, with large, irregularly shaped ooliths. Lamellibranchs and gastropods are common, usually as casts; the echinoids are all irregular. The Coral Rag is in places a hard compact limestone, with many lenticular colonies of reef-building corals. The characteristic fossils include thick-shelled forms, both of regular echinoids and lamellibranchs. The coral colonies are usually a few inches thick, and from 2 to 3 ft. in diameter; they are often separated and surrounded by oolitic limestones in which thin-shelled lamellibranchs, gastropods, and brachiopods occur. Attempts have been made to explain the position of the various exposures of the Rag and Oolite by hypotheses involving folding and faulting, but it is at least equally likely that these facies interdigitate.

### *KIMERIDGE CLAY*

The *Kimeridge Clay* was described in detail by Roberts in 1892; Kitchen and Pringle have since classified some of the horizons in terms of a more modern zoning, and their results, summarised by Dr Arkell, are given in the right-hand column of the following correlation table:

| Zones (Roberts, 1892) | Lithology | |
|---|---|---|
| *Discina latissima* [*Orbiculoidea*] | 7 ft. papery shale | |
| *Exogyra virgula* | 3 ft. grey-black shale | *Aulacostephanus* |
| *Ammonites alternans* | 24 ft. clays with sandy and papery shales | |
| *Astarte supracorallina* [*extensa*] | 9 in. fissile sandy clay | *mutabilis* |

This succession is seen in the Roslyn (Roswell) Pit at Ely, where a band of large septarian nodules separates the two upper zones. Roberts considered beds exposed at Littleport to be older, and proposed a zone of *Ostrea deltoidea* [*delta*] to include the base of the Kimeridge as exposed around Haddenham. The upper part of the Kimeridge Clay is absent, the Lower Greensand resting unconformably on it. At Upware, a few feet of Kimeridge Clay (exact age unknown) wedges out under the Lower Greensand against the Corallian ridge. Lithologically, the Kimeridge Clay is distinguished from the other Jurassic Clays by the presence of paper shales. Near Southery in West Norfolk, Dr Arkell records, from borings, thicknesses of about 145 ft. of Kimeridge Clay, but the faunal succession is incomplete.

## *THE LOWER GREENSAND*

The *Lower Greensand*, largely covered by drift, extends from Gamlingay in a north-easterly direction to the Fenland, where it caps the ridges at Haddenham and Ely; outcrops are also found upon the flank of the Upware-Wicken ridge. It is an important water-bearing formation. The best exposure is to the south-west of the County boundary at Sandy (Beds), where strongly false-bedded, yellow sands occur; and, about 100 ft. above the base, a pebble bed has been recorded containing derived ammonites (*Pavlovia* spp.) characteristic of the Hartwell Clay. At Upware, the Lower Greensand is represented by about 12 ft. of yellow sand, with a basal conglomerate with layers of phosphatic nodules; this oversteps the Kimeridge Clay on to the Corallian. The rare Aptian ammonites at Upware include phosphatised fine-grained internal casts, which Keeping suggested were derived; they are identified by Dr Spath as species of *Deshayesites*, characteristic of the upper Lower Aptian. There are also a few forms, with the shell preserved round a sandstone matrix, which are possibly indigenous. Unfortunately these cannot be determined with certainty, but Dr Spath compares them with *Columbiceras* which is said to occur elsewhere in the lower Upper Aptian. A badly preserved internal cast of ?*Tropaeum* sp., identified by Keeping as *Ancyloceras hillsi*, contains a similar matrix.

Keeping attempted to identify the various derived fossils found in the Lower Greensand of Upware, Potton and Sandy, and he claimed that Wealden, Neocomian, Portlandian, Kimeridgian, Corallian and Oxfordian forms are to be found. His identifications, especially those of the ammonites, are, however, in need of revision. Since the Lower Greensand rests in different places on Oxford Clay, Ampthill Clay, the Upware Corallian and Kimeridge Clay, fossils derived from these formations are to be expected. A Carboniferous trilobite has been found at Sandy, and fossils identified as of Lower Palaeozoic age at Potton (Beds).

The indigenous fossils at Upware included abundant porifera, polyzoa, lamellibranchs, and especially brachiopods. The brachiopods of Brickhill at Upware were the subject of a detailed study by Keeping, who in 1883, from examination of over 15,000 specimens, arranged the species in three morphological series, which are preserved in the Sedgwick Museum.

### *GAULT*

The *Gault* is summarised in the table below, based on the work (some unpublished) of Dr Spath, whose identifications of the ammonoid faunas of Cambridgeshire have established their horizons with greater precision than was hitherto possible.

| | Age | Zone | Beds at Folke-stone | Exposures in Cambridgeshire |
|---|---|---|---|---|
| Upper Gault | Pleurohoplitan | *dispar* | | |
| | | *substuderi* | XIII | [Cambridge Greensand] |
| | Pervinquierian | *aequatorialis* | XII | |
| | | *auritus* | XI | Burwell, Barnwell |
| | | *varicosum* | X | Wicken, Barnwell |
| | | *orbignyi* | IX | Wicken |
| Lower Gault | Dipoloceratan | *cristatum* to *intermedius* | I–VIII | Not known |
| | Hoplitan | *dentatus* | I | Upware, Oakington |
| | | *benettianus* to *mammillatus* | — | Not known |

The Lower Gault is exposed, although very badly, at Upware, where a fairly large ammonite fauna typical of the *dentatus*-zone has been found; the beds are a dull tenaceous clay, but the lowest layers are glauconitic and sandy. A boring at Chapel Lane, Wicken passed through 59 ft. of Gault Clay; the top 18 ft. were calcareous clay, below which was a bed 5 ft. thick with ammonites characteristic of the *orbignyi-varicosum* zones. Ammonites were not recorded in the lower 36 ft., the base of which was sandy with a few pebbles. This section illustrates a point which must be borne in mind when reading the above table. These records are based entirely on ammonites, which are rarely found throughout the whole of a section. Further, the Gault is badly exposed. Lack of records from the upper Lower Gault, for instance, may possibly be due to non-exposure. Ammonites typical of the base of the Upper Gault have been found at Landbeach;

those of the lower part of the Upper Gault in an ice-transported boulder at Ely; and those of the *auritus*-zone in a pit one mile west of Burwell. The section at Barnwell revealed over 50 ft. of grey and blue calcareous clays, with an inconstant limestone band near the bottom. The thickness of the Gault varies between 100 ft. and 200 ft.; to the north-east it is reduced to 60 ft. at Methwold, Norfolk. Here, the rock passes laterally into a white calcareous clay, and finally into Red Chalk at Hunstanton on the Norfolk coast.

### *CAMBRIDGE GREENSAND*

The *Cambridge Greensand* is a thin bed of calcareous clay with glauconite grains and phosphatic nodules, resting on a well-defined surface of Gault Clay, but passing gradually into the Chalk Marl above. The glauconitic grains may be foraminiferal casts; they are commonest at the base, and disappear as they are traced upwards into the Chalk Marl. The phosphatic nodules occur usually within a layer about one foot thick. They are sometimes very rare, and are commonest in the deeper depressions in the Gault surface. They, too, become fewer and smaller when traced upwards, and usually disappear earlier than the glauconitic grains. The nodules are often black or dark brown, but there are light brown examples, and all intermediate stages are to be found. Many include fossils or are internal casts of fossils; often they have adherent lamellibranchs (such as *Dimyodon nilssoni*). More rare, are pebbles of igneous and sedimentary rocks; similar pebbles have been recorded throughout the English Chalk, but are sporadic. The sudden change of lithology between the Gault and Cambridge Greensand makes the junction quite distinct. The upper surface of the Gault is irregular, and is obviously an erosion surface; small irregular tubes filled with a matrix of the Cambridge Greensand penetrate downwards into it. The passage from Cambridge Greensand into Chalk Marl is gradual, so that the thickness of the Cambridge Greensand cannot be given accurately; it is, however, approximately one foot thick, but may be more in the deeper hollows of the Gault. Lithologically, the Cambridge Greensand is the basal pebble-bed of the Chalk Marl.

The fauna recorded from the Cambridge Greensand is very large, partly because it was once extensively worked for its phosphatic nodules and therefore offered abundant opportunities to collectors, partly because the phosphatisation of its fossils increased their chances of survival. Some of the species, such as *Terebratulina triangularis*, are always unphosphatised; but the majority are internal phosphatised casts. Dr Spath considers that all the ammonites may have come from the *aequatorialis*- to *substuderi*-zones of the Upper Gault.

### *THE CHALK*

The *Chalk Marl*, about 80 ft. thick, is a compact bluish argillaceous limestone, weathering brown. Gasteropod and ammonoid casts are characteristic. It is exposed at the Norman Cement Works, just south-east of Cambridge, and near Barrington.

The *Burwell Rock* (15 ft.–20 ft.), seen at Burwell, is well jointed, brownish in colour, and contains small brown phosphatic nodules, and an abnormal proportion of small shell-fragments. It passes upwards into the *Grey Chalk* (70 ft.), which is distinguished by curvilinear jointing, often almost horizontal and simulating bedding. *Holaster subglobosus* is common in the lower part, but is replaced by *H. gregoryi* in the upper (seen in the pit on the Golf Course on the Gogs). At the top is a variable series with yellowish laminated marl seams in which *Actinocamax plenus* reaches its greatest size: this is the attenuated representative of the Belemnite Marls.

The subdivisions of the Chalk are summarised in the following table:

| | Zone | Rock-bands and Lithology |
|---|---|---|
| Upper Chalk | *Micraster coranguinum* | White Chalk with flints |
| | *Micraster cortestudinarium* | |
| | | Top Rock |
| | *Sternotaxis planus* | Chalk with flints |
| | | Chalk Rock |
| Middle Chalk | *Terebratulina lata* | Chalk with flints |
| | *Inoceramus labiatus* | White Chalk |
| | | Melbourn Rock |
| Lower Chalk | *Holaster subglobosus* | Belemnite Marls |
| | | Grey Chalk |
| | | Totternhoe Stone or Burwell Rock |
| | *Schloenbachia varians* | Chalk Marl (passing down into Cambridge Greensand) |

The *Melbourn Rock* (about 10 ft. thick) is a hard limestone; on weathered surfaces it is seen to be nodular. Greenish marl seams occur, and sometimes the marl wraps round isolated nodules. *Inoceramus labiatus*, *Rhynchonella cuvieri* and *Discoidea dixoni* are characteristic. Above the Melbourn Rock is white chalk with some nodular bands. The junction between the *labiatus*- and *Terebratulina*-zones is difficult to define, partly owing to lack of exposures; the total thickness of the Middle Chalk, however, is about 200 ft., the upper part consisting of white chalk with marl seams and flints.[1] Characteristic fossils of the *Terebratulina*-zone are *Terebratulina lata*, *Sternotaxis planus*, *Micraster corbovis* and a variety of *Echinocorys scutatus*, a species elsewhere diagnostic of the Upper Chalk. There are exposures near Dullingham Station, at the Linton Whiting Works, and east of Great Chesterford.

As in the Thetford district, the *Chalk Rock* of the Cambridge district occurs above the base of the Upper Chalk. At the best exposure (Underwood Hall) about 6 ft. of white blocky chalk with tabular flints underlies the Chalk Rock, which consists of hard patches of chalk embedded in soft white chalk, often with no clear-cut demarcation. Green-coated nodules are absent. Ammonoids and gasteropods are common (they are almost absent between the Chalk Rock and the Chalk Marl), and a typical *Hyphantoceras reussianum* fauna occurs. The lithology of the Chalk Rock reappears towards the top of the *planus*-zone in the *Top Rock*. This is distinguished from the Chalk Rock by the occurrence at its upper surface of a hard limestone with pinkish brown nodules, about one foot thick, with a definite top crowded with green-coated nodules, some of which are internal casts of *Micraster cortestudinarium*. The best exposure of the Top Rock is near Westley Waterless; it can also be seen south of Higham in West Suffolk.

The *Micraster*-zones of the Chalk are largely concealed under Boulder Clay; the *cortestudinarium*-zone is seen north-west of West Wratting, and the *coranguinum*-zone near Shudy Camps and Saffron Walden. The Chalk is white and flints are plentiful.

## THE PLEISTOCENE DEPOSITS OF THE CAMBRIDGE DISTRICT[2]

In Cambridgeshire no exposures have been found of beds comparable in age with the Crag and Early Pleistocene deposits of the eastern parts of Norfolk and Suffolk. Coarse gravels and sands, fan-wash fingering out from the *Lower Chalky Boulder Clay* ice sheet, can be seen underlying that Boulder Clay on the high ground south-east of Cambridge—on the Gogmagog Hills and at Haverhill in Suffolk. The ice sheet advanced over a fairly deeply dissected landscape as far south as the London Basin.

[1] H. Dixon Hewitt has shown that part at least of the famous Brandon Flint Series (Suffolk) belongs to the *Terebratulina*-zone.

[2] By T. T. Paterson, M.A.

Rubbling and thrusting of the Chalk took place where the ice met the Chalk escarpment, and the consequent structures are well exposed south of Royston and on Chalk Hill, north-east of Newmarket. The Boulder Clay is generally confined to the high ground and upper slopes. It is characterised by a blue colour, and by erratics of chalk, Lincolnshire flint, Yorkshire sandstones and coals, and Scottish quartz dolerites, quartzites and granites, as well as by rocks of Scandinavian origin.

Upon the retreat of the ice, the land, several hundred feet higher than to-day, was strongly eroded. Deep, steep-walled valleys were formed, and these were subsequently filled up either by glacial drift or during a period

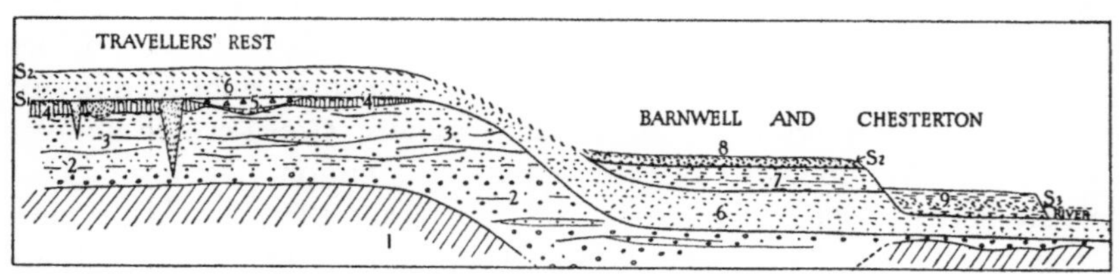


Fig. 5.

Diagrammatic Composite Section of the Terraces around Cambridge.

1. Gault with surface rucked by sludging.
2. Lower even-bedded series of Travellers' Rest Pit, with erratics derived from Lower Chalky Boulder Clay.
3. Uneven-bedded series of Travellers' Rest Pit, with rolled Lower Palaeolithic tools and cold fauna.
4. Loess-loam.
5. Upper Chalky Boulder Clay in lenses.

$S_1$ Solifluxion band with frost cracks and polygonal soil forms of Upper Chalky Boulder Clay age.

6. Succeeding interglacial aggradation gravels.
7. Middle Terrace gravels with warm fauna and Late Clacton-Levallois-Acheul industry.
8. Loams and gravels with cold fauna and associated solifluxion band ($S_2$).
9. Lower Terrace; fine gravel and silt with poorly marked solifluxion band ($S_3$).

of rapid aggradation and change of base level. Conglomerates and coarse gravels were deposited on the north of the main Chalk escarpment between bosses of Chalk (Barnham), Boulder Clay and outwash fans; and the finer facies were laid down as a flattened spread in the Fen region, determining the essential features of the present-day landscape (Shrubhill, near Feltwell). A warm fauna has been found in a gravel of a late stage of aggradation (Fakenham), and it is probable that the Barrington gravels are of this age. During a subsequent wet period the surface of the gravels carrying the warm fauna was sludged, and, on this sludged surface, during a drier time, brown loess-like loams accumulated (Brandon; Travellers' Rest Pit, Cambridge).

A few rolled Early and Middle Acheulean bifaces and Early Clactonian flakes have been found in the warm gravels; while *in situ* at the top and on the surface, there is a series of Middle Clactonian industries. In the brown loams, a Middle to Upper Acheulean industry occurs in many localities.

The brown loam is capped with outwash gravels heralding the onset of the *Upper Chalky Boulder Clay* ice sheet which advanced along the valleys, partially enveloping the hill slopes and penetrating almost to the London Basin at Hertford. This Boulder Clay is distinct from the earlier deposit in its brown colour, due to the included brown interglacial loam and to the large quantity of Bunter erratics. The earlier glacial and interglacial deposits are folded and overthrust. There were two advances of the ice and, during the intra-glacial period, some sands and loess-loams were deposited in isolated pools on the surface of the Boulder Clay of the first advance; there is an exposure at West Stow (Suffolk). It is probable that the High Lodge Late Clactonian industry belongs to this period, because, along with earlier forms, tools of that age appear in the gravels of Warren Hill which is an outwash deposit formed during a halt in the retreat of the ice sheet. Decalcification of Chalky Boulder Clay has given rise to a great part of the sands of the Breckland.

The three *terraces* of the river gravels are composite, and are cut out of the earliest interglacial gravels and Upper Chalky Boulder Clay. After the deposition of the latter, rapid erosion cut channels over 30 ft. deep. Subsequent aggradation filled these, and this gravel filling caps the Upper Terrace. During the succeeding interglacial phase, 20 to 30 ft. channels were cut and the Middle Terrace was formed of fine gravel carrying a warm fauna and an industry with the very latest Clactonian technique conjoined with Levallois and Acheulean (St Neots; Milton Road, Cambridge). A solifluxion layer on these gravels of the Upper and Middle Terraces, and a deposit with a cold flora, indicate a third cold period (Barnwell and Chesterton). The Lower Terrace is cut out of all the preceding deposits and is composed of fine gravel and silt with a poorly marked solifluxion level in the gravel.

## THE POST-GLACIAL DEPOSITS OF FENLAND[1]

The post-glacial deposits of the Fenland occupy a shallow basin centring upon the Wash. They are of two types. On the landward side, they are composed largely or wholly of peat, which has formed as a result of an accumulation of fresh water augmented by the numerous large river

[1] By H. Godwin, M.A., Ph.D.

systems that enter the fens from the surrounding upland, and been maintained by poor drainage gradients towards the sea. On the seaward side, the fenland deposits are silts and clays laid down under conditions of greater or less salinity (see Fig. 47). Alterations in the former relative levels of land and sea have left their trace in the disposition of the two types of deposit. During a phase of marine transgression the silts and clays extended inland above the peat beds, and in the ensuing phase of regression, peat extended seawards over the silts and clays. In this way, the silts and clays from the seaward side, and the peats from the landward side, interdigitate with one another.

A much simplified scheme showing the relations of the chief fen-beds to one another is shown in Fig. 6. The history of formation of the fen deposits can be very briefly outlined as follows:

*The Pre-Boreal Period.* Say before 7500 B.C., a period of sparse birch-pine. Peat was forming on the present floor of the North Sea, the coast of which is now about 200 ft. below its former level. No deposits of this age are yet known from the Fenland.

*The Boreal Period.* From about 7500 to 5500 B.C., a period of birch-pine woods, but with oak and elm and hazel in increasing importance. In the deep river valleys of the Fenland peat-formation began during this period (e.g. in Little Ouse Valley), and at a few sites with local water supply. During this time the North Sea reached most of its present extent, but it did not directly affect the Fenland.

*The Atlantic Period.* From about 5500 to 2000 B.C., a warm wet climatic period marked by the sudden onset and subsequent importance of the alder and, to a smaller extent, of the lime.

At the beginning of this period, Mesolithic (Tardenois) man occupied local sand hills in the fens, but, very soon, peat-formation became widespread throughout the fen basin, and thick beds of fen sedge-peat and fen brushwood-peat were formed. Towards the end of this time the peat surface dried and became increasingly wood-covered, partly due, no doubt, to marine recession and partly to climatic dryness. There is some trace of Neolithic man at this time.

*The Sub-Boreal Period.* About 2000 to 500 B.C. At the end of the Neolithic or in the early part of the Bronze Age, an extensive but shallow marine invasion caused silt and clay to spread far inland over peat. Foraminifera and diatom analyses suggest shallow brackish lagoon conditions; this was the stage of formation of the Fen Clay.

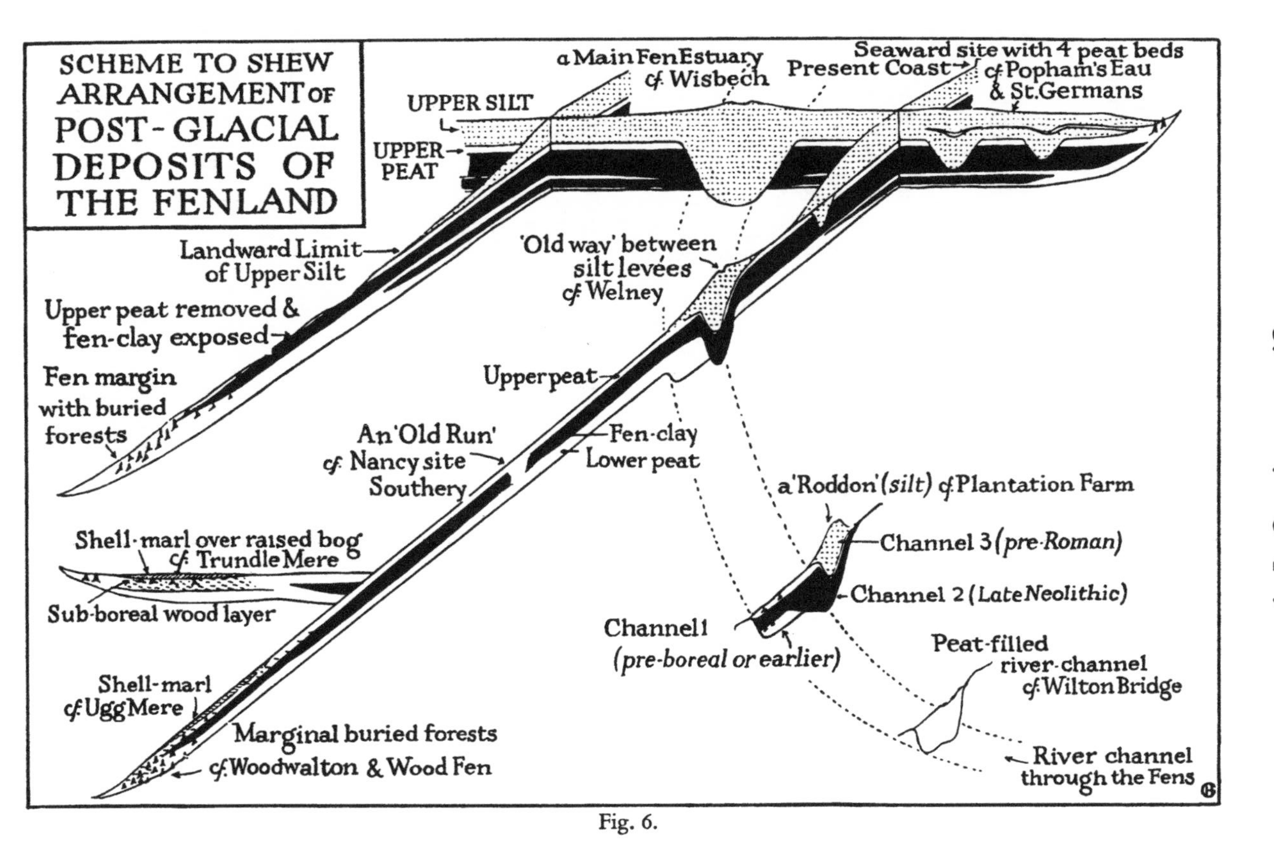
SCHEME TO SHEW ARRANGEMENT OF POST-GLACIAL DEPOSITS OF THE FENLAND
UPPER SILT
UPPER PEAT
a Main Fen Estuary cf. Wisbech
Present Coast
Seaward site with 4 peat beds cf. Popham's Eau & St. Germans
Landward Limit of Upper Silt
Upper peat removed & fen-clay exposed
Fen margin with buried forests
'Old way' between silt levées cf. Welney
Upperpeat
An 'Old Run' cf. Nancy site Southery
Fen-clay
Lower peat
a 'Roddon' (silt) cf. Plantation Farm
Channel 3 (pre-Roman)
Channel 2 (Late Neolithic)
Channel 1 (pre-boreal or earlier)
Shell-marl over raised bog cf. Trundle Mere
Sub-boreal wood layer
Peat-filled river-channel cf. Wilton Bridge
Shell-marl cf. Ugg Mere
Marginal buried forests cf. Woodwalton & Wood Fen
River channel through the Fens

Fig. 6.

In the middle and later phases of the Sub-Boreal period, semi-marine conditions were replaced by those of peat-formation. Locally, pine and birch woods developed (e.g. Wood Fen at Ely, and Woodwalton), and the fen surfaces became dry enough to have encouraged dense occupation by Bronze Age man. Along the fen margins, freedom from flooding allowed sphagnum peat to form incipient raised bogs.

*The Sub-Atlantic Period.* From 500 B.C. This period is generally recognised as colder and wetter than the Sub-Boreal. At this time the Fenland became very inhospitable, and was, apparently, shunned by Iron Age man. The period is marked by two stratigraphical events. In the peat fens were formed the shallow lakes which persisted into the last century (e.g. Whittlesey Mere, Ugg Mere). Their sites are still recognisable by the deposits of lake marl that formed on their beds. The seaward side of the fens was built up during the Roman times by the deposition of fine silt above the upper peat. These silts formed a broad belt round the Wash and they extended to high-tide level. They were densely occupied in Romano-British times,[1] especially in the last stages of their formation, and this is true also of the raised banks of the tidal rivers which form landward extensions of the silt country. These levées are now recognisable as "roddons", raised banks standing above the peat fens. They become increasingly evident as drainage causes the wasting of peat which formed over their flanks after the Roman period. The courses of the extinct waterways of the Fenland have been mapped by Major Gordon Fowler, and Fig. 7 summarises the available information about their extent and distribution. They are mostly of Romano-British age.

## THE CHALK WATER TABLE SOUTH-EAST OF CAMBRIDGE[2]

The Fenland and much of Cambridgeshire which lies below the 50 ft. contour line are not well adapted for a water-table survey. On the Chalk uplands in the south-east of the County, however, a considerable amount of work has been done since 1935 by undergraduates of the University Department of Geography. The water levels in about 120 wells, within the region covered by Fig. 8, have been measured three times a year; and fortnightly observations have been taken (during term) at about fifteen reliable and widely distributed wells. Previous to 1929, similar measurements had been made by Prof. W. B. R. King and also by officers of the Royal Engineers.[3]

[1] See p. 92 below. [2] By W. G. V. Balchin, B.A., and W. V. Lewis, M.A.

[3] See H. J. O. White, "The Geology of the Country near Saffron Walden" (*Mem. Geol. Surv.* 1932), p. 109.

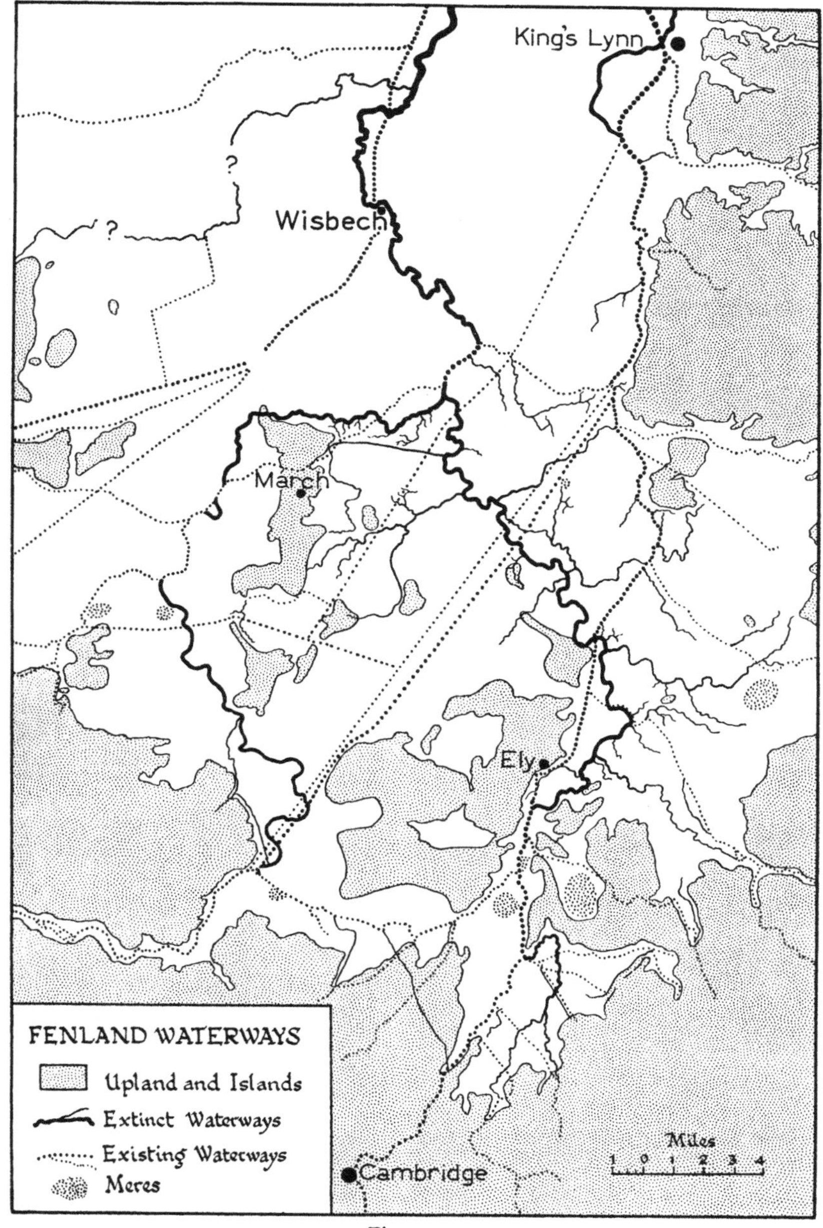


Fig. 7.

Based upon (1) Gordon Fowler, "The Extinct Waterways of the Fens", *Geog. Jour.* lxxxiii, 32 (1934); (2) additional information supplied personally by Major Fowler.

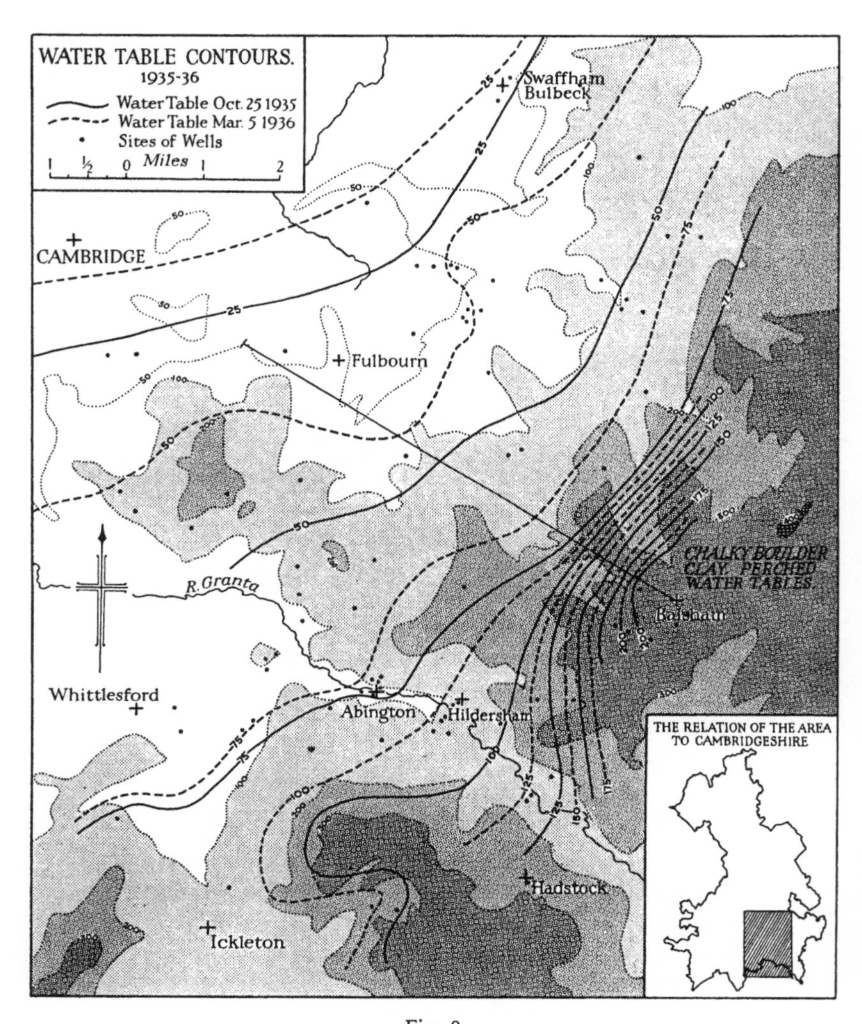


Fig. 8.

Relief Map of south-east Cambridgeshire showing Water Table contours subsequent to the drought of 1934–35. The autumn minimum and spring maximum levels are shown.

The post-1935 work has enabled water-table contours to be constructed for three different dates in each year (October, March and June), and Fig. 8 is an example of the results obtained. The October level of 1935 was selected as it came after the drought of 1934–35 and represents, therefore, an exceptionally low level. The March level of 1936 represents the spring maximum of that year, and although this was lower than the maxima for 1937 and 1938, the two readings chosen represent the greatest range within any of the three years. Although the best wells have been levelled from bench-marks, too great reliance should not be placed on the details of the contours. The errors inevitable with a party of nearly eighty students are partly counterbalanced by the large number of observations taken; but

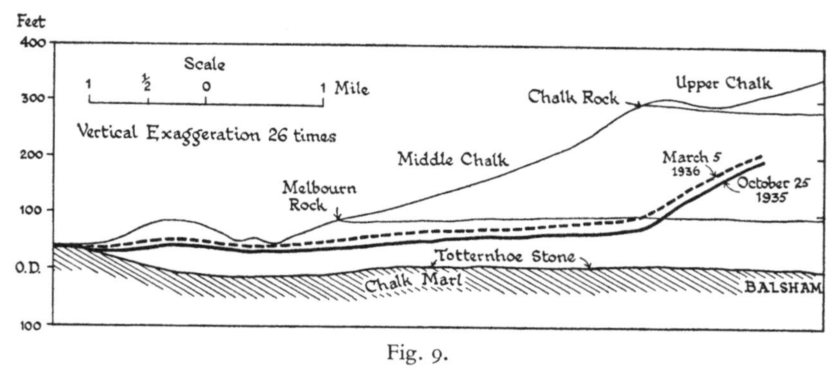


Fig. 9.

Water Table: Section along a line running north-west from Balsham (see Fig. 8).

on the other hand, the wells are irregularly distributed, and pumping, together with the recent closing of many wells, prevent great accuracy. This applies particularly to the neighbourhood of the town of Cambridge. A feature of note is the fall of the water table below the level of the Granta near Abington and Linton during October 1935; an impervious bed of alluvium presumably enabled the river to flow perched above the neighbouring Chalk water table.

The section (Fig. 9) covers the same period as the map and shows the geological formations that influence the water table. The relation between surface relief and the water table is clear, particularly in the sudden rise in level north-west of Balsham. The seasonal range is also seen to increase with the depth of the water table. The section is terminated on the west at the outcrop of the Totternhoe Stone near Cambridge, and on the east at the occurrence of several perched water tables in the Boulder Clay near Balsham. In this latter area, wells, within 40 yards of each other, differ by more than a hundred feet in their water levels.

Fig. 10 shows (1) representative rainfall figures for the county, and (2) the fluctuations in a shallow well at Great Abington. The low rainfall for the spring and summer of 1935 is reflected in the exceptionally low autumn water table. Further, a strong seasonal rhythm can be seen in the water table which is not evident in the rainfall. It is clear, particularly in 1936, that summer rainfall has little or no influence on the water table. On the

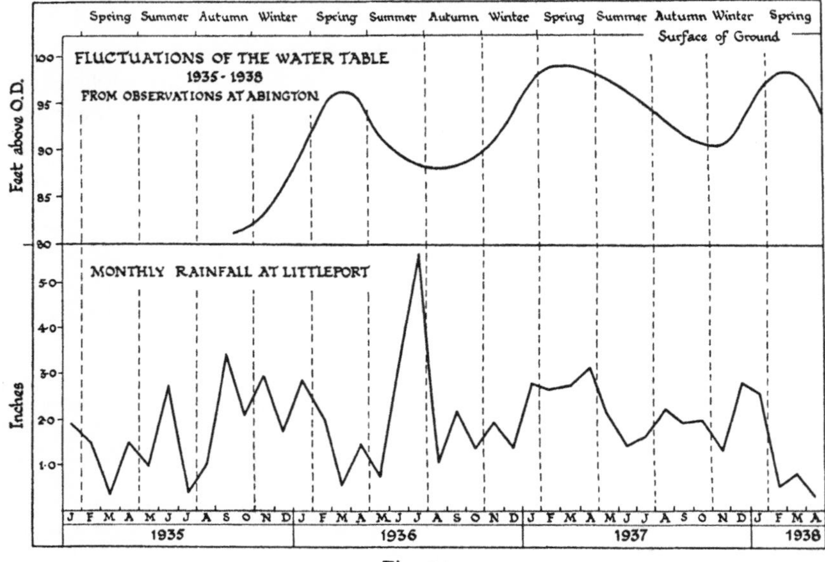


Fig. 10.

Fluctuations of water level at Great Abington in relation to the rainfall of the County, 1935–38. The rainfall for the period 1935–38 was abnormal in amount and distribution through the year. Fig. 13 below gives more representative figures for the County.

other hand, the heavy autumn and winter rainfall of 1935 probably accounts for the rapid rise in level after the drought, while the heavy winter and spring rainfall of 1936–37 explains the high and extended 1937 maximum. The drought at the beginning of 1938 is also reflected in an unusually early lowering of the water table.

## CHAPTER TWO

# THE SOILS OF CAMBRIDGESHIRE

## By H. H. Nicholson, M.A., and F. Hanley, M.A.

WITHIN THE COMPARATIVELY LIMITED AREA OF Cambridgeshire, circumstances have combined to produce a large variety of soil conditions.[1] To begin with, the Cretaceous and Upper Jurassic outcrops from the Upper Chalk down to the Oxford Clay provide a goodly range of parent materials, and superimposed upon these are large areas of drift deposits of all sorts, from recent Alluvium to Glacial Gravel (see Figs. 4 and 29). Some of these drift deposits are so thin and diffuse as to escape notice in the Drift Maps but they are of considerable importance from the soil point of view. Furthermore, the topography of the area varies from the comparatively high hills of the Chalk escarpment to the flat expanses of Fenland, so that every type of drainage condition is encountered. The types of soil parent material and the varying drainage conditions are summarised in the adjoining table.

While this table summarises drainage conditions from the point of view of soil formation and leaching, it does not fully depict land-drainage conditions. The greater part of the Fenland is dependent on artificial drainage and the use of pumps to keep the water table at a reasonable level. Outside the fen area, the upper rivers, Cam, Granta, and Rhee, are bordered by belts of gravel and alluvium which, because of the higher and more permeable areas flanking them, are characterised by high water tables and liability to flooding. The area of Gault and Chalk Marl in the Rhee Valley is peculiarly circumstanced for drainage. An outlying ridge of chalk, north of Barrington, makes this almost a land-locked basin which, although of an altitude 70 ft. or more, is afflicted with drainage conditions not unlike those of the Fenland itself. With the exception of the Boulder Clay areas, all the clay lands of the County are both impermeable and low lying, though some of them are favoured with enough fall to make field drainage moderately simple. In the Boulder Clay country, the heavy land has the advantage of fair altitude and for the most part useful

[1] Fuller descriptions, together with analytical details and profile descriptions, are contained in Bulletin No. 98 of the Ministry of Agriculture, *The Soils of Cambridgeshire* (1936), by H. H. Nicholson and F. Hanley. Detailed information concerning the soils of fruit-growing areas in West Cambridgeshire, and in the Isle of Ely, is available in Bulletin No. 61 of the Ministry of Agriculture, *West Cambridgeshire Fruit-Growing Area* (1933), by J. F. Ward; and in Research Monograph No. 6 of the Ministry of Agriculture, *A Survey of the Soils and Fruit of the Wisbech Area* (1929), by C. Wright and J. F. Ward.

slope. The ditches are often natural watercourses, and impermeability of the formation is the only obstacle to its satisfactory drainage.

PARENT MATERIALS AND SOIL-DRAINAGE CONDITIONS

(1) WITH FREE DRAINAGE. (Parent material permeable, excess water draining to depth)

A. *Drainage Excessive.* (Considerable leaching; moisture-holding capacity poor)

| LIMESTONES | GRAVELS AND SANDS ON CHALK | SANDS |
|---|---|---|
| Upper Chalk | Plateau Gravels | Lower Greensand |
| Middle Chalk | Glacial Gravels | Breckland |
| | Taele Gravels | |

B. *Drainage Free.* (Moisture-holding power satisfactory for plant growth)

| LIMESTONES | SANDS |
|---|---|
| Lower Chalk | Lower Greensand (Mid-Cambs) |
| Coral Rag | |

(2) WITH IMPEDED DRAINAGE. (Subject to continuous or seasonal water-logging)

A. *Drainage Imperfect.* (Subsoil and parent material impermeable)

CLAYS

Boulder Clay

| | |
|---|---|
| Gault | Ampthill Clay |
| Kimeridge Clay | Oxford Clay |

B. *Drainage Impeded.* (High water table due to topography; underlying stratum impermeable, or low permeability)

| LIMESTONES | GRAVELS | ALLUVIUM |
|---|---|---|
| Chalk Marl | Valley Gravels | Peat |
| Lower Chalk (base) | Old River Gravels | Silt |
| Middle Chalk (base) | | |

Most of the soils reflect to a marked degree the characteristics of their parent materials or the geological outcrop on which they lie, but this is not invariably the case. Textures vary from the loose sands of Breckland to the heaviest Gault, Boulder Clay or Oxford Clay soils. In general, the lime status of the soils is satisfactory or good, the only serious exception being the soils derived from the Lower Greensand (especially those in the west of the County), and some of the Gravels. This is to be expected of soils formed from cretaceous rocks, or drifts originating in the neighbourhood of cretaceous exposures, or associated with rivers of calcareous waters.

*Boulder Clay* soils occupy two substantial areas, one in the south-east along the Essex border, and one in western Cambridgeshire on the Huntingdonshire border. They are buff-coloured clays or clay loams, lying on clay subsoils, containing from 1 to 10 per cent calcium carbonate, much of it as rounded chalk pebbles. The depth to the grey Chalky Boulder Clay varies considerably. In some places, it can be reached within 2 ft. of the surface, but, on the lower slopes and in the smaller valley bottoms, there is often almost this depth of colluvial heavy loam material before a buff clay subsoil is reached, and this in its turn gives place to grey unweathered clay. The drainage of these soils is naturally poor, although the two areas are in the highest parts of the County. The western area is the most impermeable. On high ground, where land is flat or only gently sloping, run-off is hindered, and water-logging or surface pools are of common occurrence in a wet winter. Smaller areas of Boulder Clay soils are also found in the "islands" of the Fens.

The *Upper Chalk* formation in Cambridgeshire is covered, for the most part, by Boulder Clay. The soils on the limited exposed areas vary from thin white or grey chalky soils to brownish grey loams, depending on the proximity of the Boulder Clay and the extent of downwash from it.

The exposure of the *Middle Chalk* occupies a big proportion of Cambridgeshire, and is covered by two main classes of soil. One is a thin grey or brownish grey chalky loam, consisting mostly of fragments of chalk, and lying directly on raw chalk. This chalky, or "whiteland", type occupies the higher slopes and summits especially along the flanks of the exposure. The other is a warm brown or reddish brown loamy sand, up to 20 in. deep, lying on the chalk. This "redland" type consists chiefly of coarse sand with only a small percentage of calcium carbonate. It fills the lower slopes and flatter areas, especially around the numerous patches of *Gravel* scattered along the outcrop. The soils on these Gravels are dissimilar only in respect of their greater content of flint pebbles. All the soils on the Middle Chalk are, with the exception of hollows and limited areas on the north-west edge of the outcrop, characterised by very free drainage. This constitutes their chief drawback, and the farmer's great difficulty is to conserve sufficient moisture in the soil to carry the crops through the growing season.

The *Lower Chalk* soils differ from those of the Middle Chalk, chiefly in that they are less deep but heavier in texture and have a narrower range of colour. They are free draining but the water table, in general, comes nearer the surface, and, in places, field drainage is necessary. Both in the Lower Chalk soils and in a few of the low-lying parts of the Middle Chalk, yellow mottling, the characteristic sign of impeded drainage, is to be found in the subsoil chalk within 18 in. of the surface.

The *Chalk Marl* soils occupy low-lying ground, much of it bordering on the Fens. Both colour and texture vary as a result of surface admixture with drift materials. The soils are generally brownish grey, marly, medium to heavy, loams lying on yellowish grey marly subsoils, often with yellow or orange mottling due to high water tables. Coprolites are plentiful in the surface soil,[1] especially in those parts of the outcrop bordering that of the Gault.

The *Gault* soils are dark brownish grey in colour, and form the heaviest of the clay soils. They lie on a buff-coloured clay subsoil, which merges into blue-grey clay with orange mottlings. The formation is impermeable, as are the soils which lie on its surface, except in so far as they are opened up by tillage or by admixture with sand and gravel from neighbouring formations. Consequently, the land tends to lie wet or water-logged in its natural condition, except during dry seasons. The soils, however, have the advantage of being calcareous (2 to 10 per cent of calcium carbonate, increasing to 30 per cent or more within 3 ft. of the surface). This calcium carbonate is not present as lumps or pebbles, but is finely disseminated through the soil and so assists the formation of good tilths.

There is not a very big outcrop of *Lower Greensand* but it is important, being associated with intensive market gardening in the western part of the County, and with fruit and flower culture in the centre of the County. The soils are rich brown, loamy sands in the west, with more mellow sandy loams to the centre. The parent material is a coarse quartz sand, highly permeable, so that the chief features of the soils are their coarse open character, very free drainage, low content of organic matter and bases, frequently acidic reaction with signs of leaching and the formation of iron-pan in the subsoils. This pan in some cases, and the proximity of underlying clay in others, gives rise to localised patches where drainage is impeded and the subsoil is mottled.

The *Kimeridge Clay* gives rise to dark grey-brown clays and heavy loams, often with a poor reserve of lime. They occupy low-lying flat areas, and their chief handicaps are their poor tilth potentialities due partly to difficulties of draining and partly to their low content of calcium carbonate.

The incidence of thin washes of drift materials gives to the other Jurassic clay soils (*Oxford Clay* and *Ampthill Clay*) a character of their own. The surface textures are varied and there are wide variations in the soil profile. They are all on the heavy side, but the top soil is frequently much lighter than the subsoil. Their poor reserve of calcium carbonate, however, is a factor against the easy production of good tilths. Low lying and adjacent to the Fenland, their lack of fall and lack of internal structure make them difficult to drain.

[1] See p. 13 above, and p. 126 below.

Scattered along the existing rivers and around the southern edge of the Fens, and constituting some of the "islands", are considerable areas of *Valley and River Gravels*, of varying constitution. They give rise to soils that are gravelly, brownish grey to grey-black in colour, and loamy sands to medium loams in texture. The soils are free draining, but many of them, through their position, have high water tables except in so far as these are lowered by field drainage. Much of the material of the Gravel deposits is calcareous, but leaching has reduced the percentage of calcium carbonate in the top soils to very low values. The ground water is calcareous, however, and seriously acid soils are rare.

The *Fen Alluvial* deposits cover about half the area of Cambridgeshire. The soils which are characteristic of Fenland are of four main kinds—peat, silt, shell marl and skirt. All these soils have been formed from materials laid down in association with a river system containing calcareous waters, a fact which has had an important result on the fertility of the soils; they are almost all rich in calcium in one form or another. The surface material, of varying constitution, is the result of the deposition of inorganic particles by the rivers and estuaries in their meanderings and frequent floodings, and of inorganic material by the growth of vegetation in swamp and marginal conditions. The variation of these main factors from time to time, and from point to point, has resulted (*a*) in a "profile" of great complexity at any point, and (*b*) in considerable variations in existing soil conditions from place to place.

The *Peat* soils are composed mostly of organic matter derived from swamp vegetation. They are black or nearly so, light, spongy, and crumbly. The organic matter contains considerable amounts of exchangeable calcium and is frequently associated with a small amount of free calcium carbonate. The depth of the peat deposit varies considerably from place to place, from a few inches to 10–15 ft.; but it is not always in an uninterrupted layer. Occasionally, as many as five separate bands have been proved, interlayered with "buttery clay", silt, or sandy deposits. Much of the peat does not lie directly on the older formations but on the characteristic clay, a dark blue-grey greasy material, entirely unlike the older clay formations. The depth of the peat[1] and the nature of the underlying material are points of major importance in determining the agricultural value of the land.[2]

The *Silt* soils are associated chiefly with areas of marine deposits of fine rounded quartz grains with small flakes of muscovite, occurring at the seaward or northern end of the County. They show a big range of texture

[1] For the shrinkage and wastage of the peat, see p. 186 below.

[2] For the practice of "claying" the peat, see pp. 120–1 and 152 below.

from place to place but are deep and uniform. The physical make-up of all these soils is peculiar, and unlike anything commonly encountered elsewhere. Coarse sand particles are generally absent, but fine sand is present in quantity, especially in the lighter silts. The silt fraction is more prominent in the heavier silts. The *Shell Marl* soils, derived from material formed in the clear water of shallow meres, occur farther south as patches among the true black peats. They are white or grey in colour, highly calcareous, with numerous small freshwater shells or shell fragments present. The chief shell marl areas are on the sites of Stretham Mere and Soham Mere. *Skirt* soils, from their location and constitution, appear to be the result of conditions where both the accumulation of peat and the deposition of silt or drift have alternately held sway. They might be described as mineral soils with a rather higher content of organic matter than usual and a darker or black colour.

Representative data of these typical soils are summarised in the following table:

CONSTITUTION OF MAIN TYPES OF TOP SOILS

The figures show the percentages of carbonates and of each of the four usual mineral-particle size-groups in the air-dry soil after passing through a 2 mm. sieve

| Type | Coarse sand | Fine sand | Silt | Clay | $CaCO_3$ |
|---|---|---|---|---|---|
| South-west Boulder Clay | 8·8 | 18·0 | 15·5 | 32·0 | 8·2 |
| South-east Boulder Clay | 18·4 | 26·0 | 11·8 | 24·7 | 10·6 |
| Middle Chalk (Redland) | 46·6 | 23·3 | 7·4 | 9·2 | 4·6 |
| Middle Chalk (Whiteland) | 22·3 | 13·4 | 6·0 | 6·7 | 38·0 |
| Lower Chalk | 32·2 | 14·1 | 8·0 | 9·5 | 27·3 |
| Chalk Marl | 21·4 | 17·7 | 10·0 | 22·5 | 23·7 |
| Gault | 3·9 | 7·4 | 17·7 | 61·1 | 7·9 |
| Lower Greensand (West Cambs) | 74·5 | 7·6 | 2·8 | 8·0 | 0·0 |
| Lower Greensand (Mid-Cambs) | 46·2 | 20·2 | 8·3 | 15·2 | 0·15 |
| Kimeridge Clay | 20·0 | 24·4 | 15·8 | 34·0 | 0·0 |
| Ampthill Clay | 22·1 | 14·9 | 8·8 | 46·7 | 0·06 |
| Old River Gravel | 54·0 | 19·0 | 8·6 | 13·1 | 0·5 |
| Valley Gravel | 45·8 | 25·5 | 10·7 | 9·8 | 0·5 |
| Fen Peat | 1·6 | 10·5 | 21·8 | 15·9 | 5·6 |
| Fen Silt (Light) | 0·8 | 64·5 | 14·4 | 7·6 | 2·6 |
| Fen Silt (Heavy) | 0·5 | 34·2 | 19·2 | 30·9 | 1·6 |

CHAPTER THREE

# THE CLIMATE OF CAMBRIDGESHIRE

## By A. S. Watt, PH.D.

THE MAIN FEATURE OF THE CLIMATE OF THE NEIGHBOURHOOD of Cambridge lies in the definite approach it makes to the continental type.[1] Its latitude and position on the western side of a continent enable it, of course, to share with other parts of the British Isles in the equable climate associated with oceanic conditions. But its position in relation to the continental mainland also allows it to share in continental characteristics. The climate of Cambridge, in fact, may be described as transitional. Just how far it departs from oceanity and how near it approaches continentality is the main theme of this chapter.

To throw into relief the essential features of the Cambridge climate comparison is made, in the account that follows, between meteorological data from selected stations lying approximately in the same latitude across Europe from west to east. Valentia, Cambridge, Berlin, and Orenburg represent a transition from the oceanic to the continental type of climate.

### TEMPERATURE

The data given in Table 1, and graphically presented in Fig. 11, show a slightly decreasing mean annual temperature from west to east.

At Valentia, the maximum falls in August, but the value is only slightly above that for July when the maximum occurs at the other three places: the minimum, on the other hand, falls in February at Valentia, but in January at the other three stations. The range of the monthly means increases from 14·8° F. at Valentia, through 22·4° F. at Cambridge, and 33·7° F. at Berlin, to 67·3° F. at Orenburg, an increase partly due to higher summer temperatures but mainly to lower winter temperatures.

The departure from oceanity, seen in the increasing range of monthly means, is emphasised by the monthly extremes of normal maxima and minima for Valentia, Cambridge, and Berlin (Table 2; Fig. 12), which show (*a*) an increasing annual range of monthly extremes (Table 2, last

[1] It is a pleasure to acknowledge my indebtedness to Dr G. C. Simpson, Director of the Meteorological Office, for placing the services of his department at my disposal. Mr E. G. Bilham of that department has been particularly kind in elucidating points of detail. His book on *The Climate of the British Isles* (1938) appeared during the preparation of this chapter and fuller information will be found there. For the use made of the data I must accept responsibility.

Table 1. *Mean annual and mean monthly Temperatures in °F.*

| | Jan. | Feb. | Mar. | Apr. | May | June | July | Aug. | Sept. | Oct. | Nov. | Dec. | Mean annual temp. |
|---|---|---|---|---|---|---|---|---|---|---|---|---|---|
| Valentia (1906–35) | 44·9 | 44·3 | 45·2 | 47·3 | 52·4 | 56·3 | 59·0 | 59·1 | 56·3 | 52·2 | 47·0 | 45·4 | 50·7 |
| Cambridge (1906–35) | 39·3 | 39·7 | 42·3 | 46·3 | 53·5 | 58·0 | 61·7 | 61·3 | 56·9 | 50·3 | 42·9 | 39·9 | 49·3 |
| Berlin | 30·7 | 32·9 | 37·8 | 45·7 | 55·8 | 62·1 | 64·4 | 62·6 | 56·8 | 47·8 | 38·8 | 33·3 | 47·4 |
| Orenburg | 4·3 | 7·7 | 18·5 | 39·2 | 58·6 | 67·5 | 71·6 | 67·5 | 55·4 | 39·6 | 23·7 | 12·2 | 38·8 |

Table 2. *Monthly extreme Temperatures in °F.*

| | Jan. | Feb. | Mar. | Apr. | May | June | July | Aug. | Sept. | Oct. | Nov. | Dec. | Range |
|---|---|---|---|---|---|---|---|---|---|---|---|---|---|
| Valentia: | | | | | | | | | | | | | |
| Max. | 53·1 | 54·0 | 55·9 | 61·0 | 68·0 | 72·0 | 71·1 | 71·1 | 69·1 | 62·1 | 57·0 | 55·0 | 18·9 |
| Min. | 28·9 | 30·0 | 30·9 | 34·0 | 37·9 | 43·0 | 46·0 | 46·0 | 42·1 | 35·1 | 32·0 | 30·0 | 17·1 |
| Cambridge: | | | | | | | | | | | | | |
| Max. | 54·0 | 55·9 | 63·0 | 69·1 | 75·0 | 81·0 | 83·8 | 82·9 | 78·1 | 68·0 | 59·0 | 55·0 | 29·8 |
| Min. | 19·9 | 21·0 | 23·0 | 26·2 | 30·0 | 37·9 | 43·0 | 42·1 | 36·0 | 28·9 | 24·1 | 21·1 | 23·1 |
| Berlin: | | | | | | | | | | | | | |
| Max. | 46·6 | 49·3 | 60·4 | 69·8 | 84·2 | 88·0 | 88·9 | 85·5 | 79·3 | 67·6 | 54·0 | 48·0 | 42·3 |
| Min. | 10·6 | 16·0 | 21·4 | 30·6 | 37·6 | 45·9 | 50·2 | 48·7 | 41·9 | 31·8 | 23·5 | 16·3 | 39·6 |

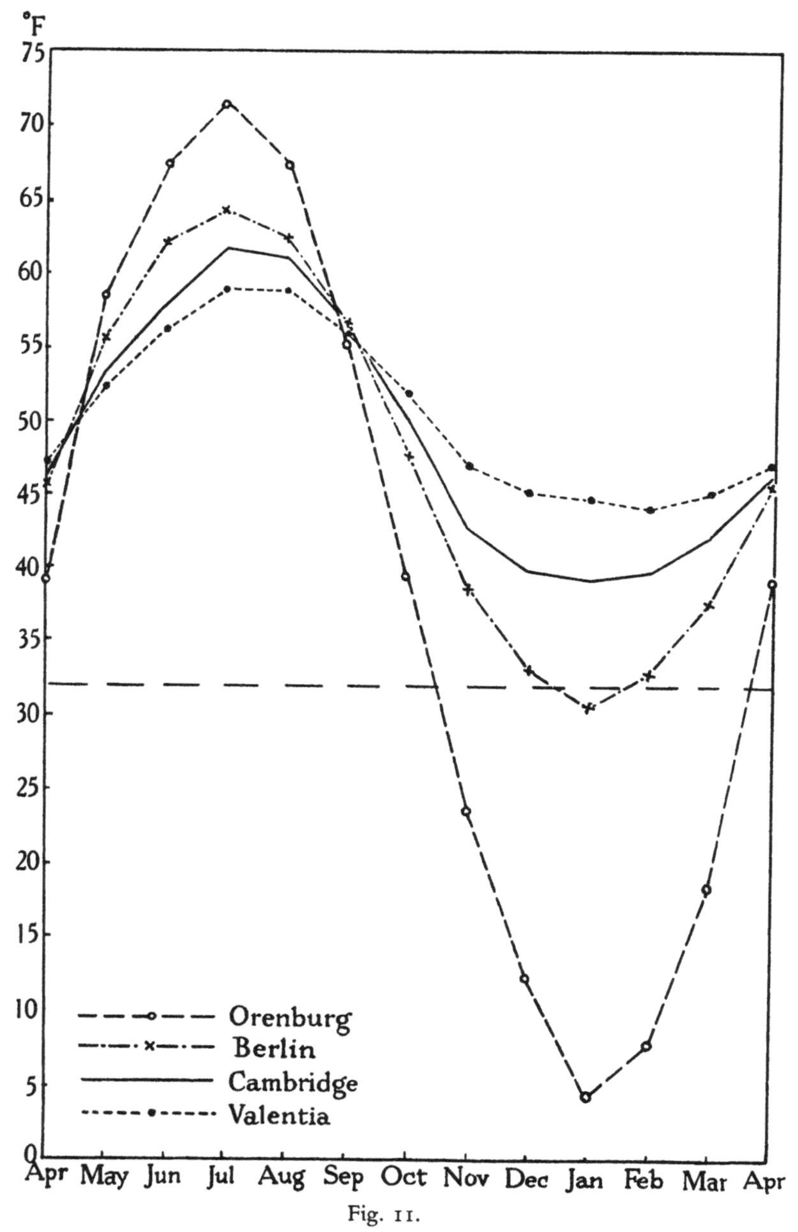


Fig. 11.

Mean monthly temperatures at Valentia, Cambridge, Berlin and Orenburg.

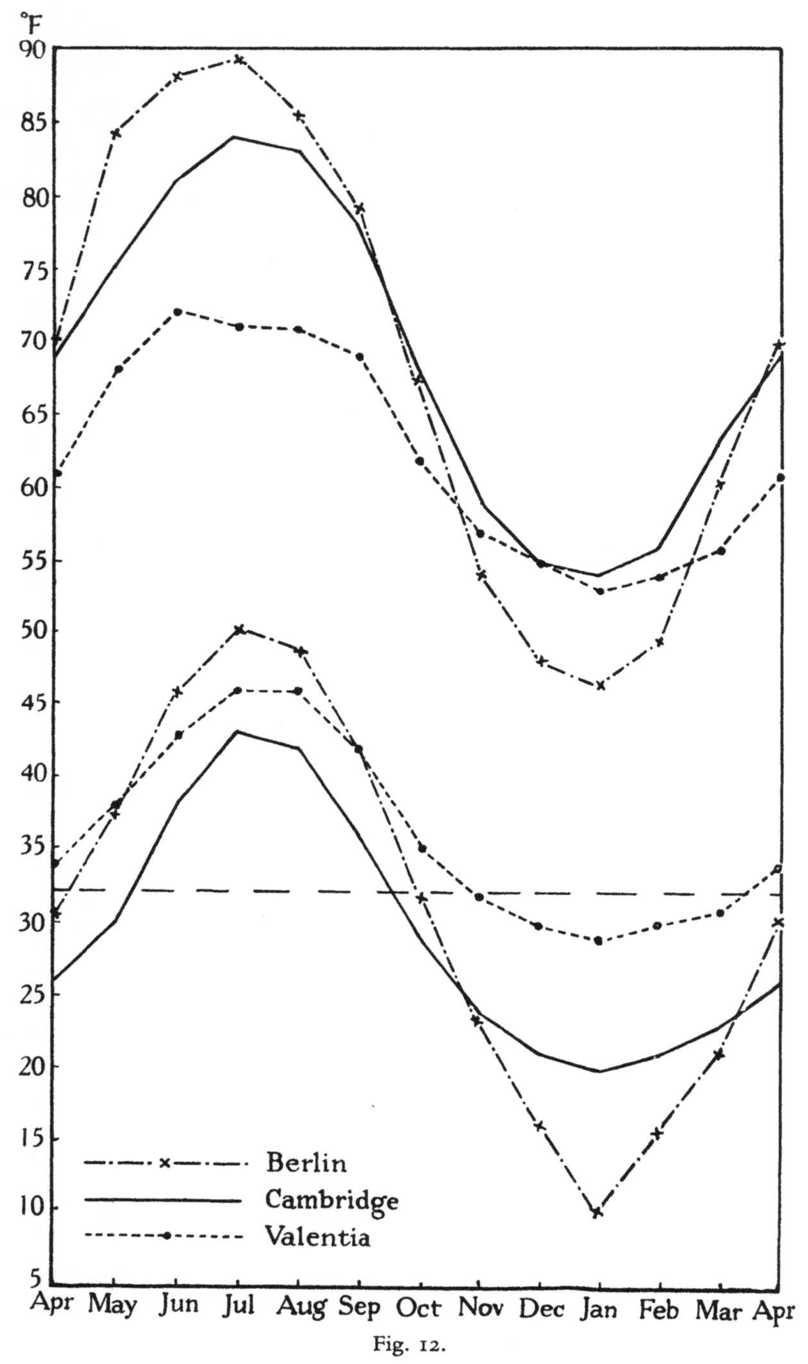


Fig. 12.

Mean monthly extremes of temperature at Valentia, Cambridge and Berlin.

column); (*b*) an increasing range of temperature in any month; and (*c*) increasing differences between the summer maxima and between the winter minima. The values for the winter maxima and summer minima approximate. But Cambridge is peculiar in having the highest winter maxima and the lowest summer minima of all three stations.

These low summer minima around Cambridge are particularly important, for they indicate the frequency of frosts (Table 3). Extreme minima below 32° F. are recorded for Valentia from December to March; for Berlin from October to April; but for Cambridge from October to May. Indeed, at Cambridge serious frosts quite often occur in the beginning of June and the only month really free from frost is July. Winter frosts also are severe, and the damage they do is likely to be increased because snow affords an efficient protection only on a few days of the year: in an average year snow lies in the morning for 12 days only. Skating is enjoyed every other year, and the last occasion when the Cam was converted into a highway was February 1929.

## RAINFALL

Both in the amount of rainfall and in its distribution throughout the year, Cambridge again shows a distinct approach to the continental type of climate (Tables 4 and 5; Fig. 13).

Like the continental stations, Cambridge has a low yearly total: Valentia has over twice as much rain as Cambridge, and its lowest monthly value is much greater than the highest at Cambridge.

The total for Cambridge itself is a fair sample of the annual rainfall of the County: the average for twenty-eight stations within the County, at altitudes varying from 6 ft. O.D. to 286 ft. O.D., is 22·28 in. The range is narrow—from 20·6 in. at Upwell to 24·7 in. at Conington. To the west, in Huntingdon and Bedford, and to the east, in Suffolk, the rainfall is slightly higher.

The distribution of the rainfall throughout the year is particularly interesting. The typical oceanic climate shows a summer minimum and a winter maximum, the typical continental the reverse. Throughout the greater part of the British Isles, the rainfall of the winter half of the year is greater than that of the summer half, a feature which the country shares with a strip of Atlantic seaboard of the continent. But a relatively small area in east-central England, including Cambridge, shows the reverse, namely, a greater fall during the summer half than during the winter half year. This is a continental feature that is further emphasised by a consideration of the details. A glance at the graph for Orenburg shows the typical feature of a continental climate, namely, the high peak in late

Table 3. *Normal number of days with ground frost for each calendar month*

| | Jan. | Feb. | Mar. | Apr. | May | June | July | Aug. | Sept. | Oct. | Nov. | Dec. | Year |
|---|---|---|---|---|---|---|---|---|---|---|---|---|---|
| Cambridge (1908–20) | 18·6 | 17·5 | 18·4 | 12·8 | 4·1 | 0·6 | 0·0 | 0·0 | 2·3 | 7·4 | 14·7 | 15·5 | 111·9 |
| Falmouth (1914–20) | 7·6 | 8·3 | 10·0 | 5·4 | 0·1 | 0·0 | 0·0 | 0·0 | 0·3 | 1·3 | 6·2 | 9·1 | 48·3 |

Table 4. *Annual and monthly rainfall in inches*

| | Jan. | Feb. | Mar. | Apr. | May | June | July | Aug. | Sept. | Oct. | Nov. | Dec. | Year |
|---|---|---|---|---|---|---|---|---|---|---|---|---|---|
| Valentia (1885–1915) | 5·49 | 5·20 | 4·54 | 3·67 | 3·17 | 3·20 | 3·78 | 4·79 | 4·14 | 5·57 | 5·46 | 6·64 | 55·65 |
| Cambridge (1885–1915) | 1·50 | 1·28 | 1·47 | 1·36 | 1·76 | 2·11 | 2·16 | 2·35 | 1·61 | 2·36 | 1·93 | 1·93 | 21·82 |
| Berlin | 1·89 | 1·34 | 1·50 | 1·65 | 1·93 | 2·17 | 2·99 | 2·21 | 2·01 | 1·54 | 1·50 | 2·05 | 22·78 |
| Orenburg | 1·1 | 0·8 | 1·0 | 0·9 | 1·4 | 2·0 | 1·7 | 1·3 | 1·3 | 1·2 | 1·2 | 1·2 | 15·1 |
| Wisbech | 1·68 | 1·39 | 1·60 | 1·32 | 1·64 | 1·99 | 2·27 | 2·42 | 1·91 | 2·66 | 2·06 | 2·05 | 22·99 |
| March | 1·60 | 1·29 | 1·58 | 1·32 | 1·73 | 1·97 | 2·37 | 2·39 | 1·80 | 2·60 | 2·05 | 2·11 | 22·81 |
| Stretham Engine | 1·34 | 1·10 | 1·27 | 1·19 | 1·73 | 2·01 | 2·45 | 2·42 | 1·71 | 2·33 | 1·83 | 1·82 | 21·20 |

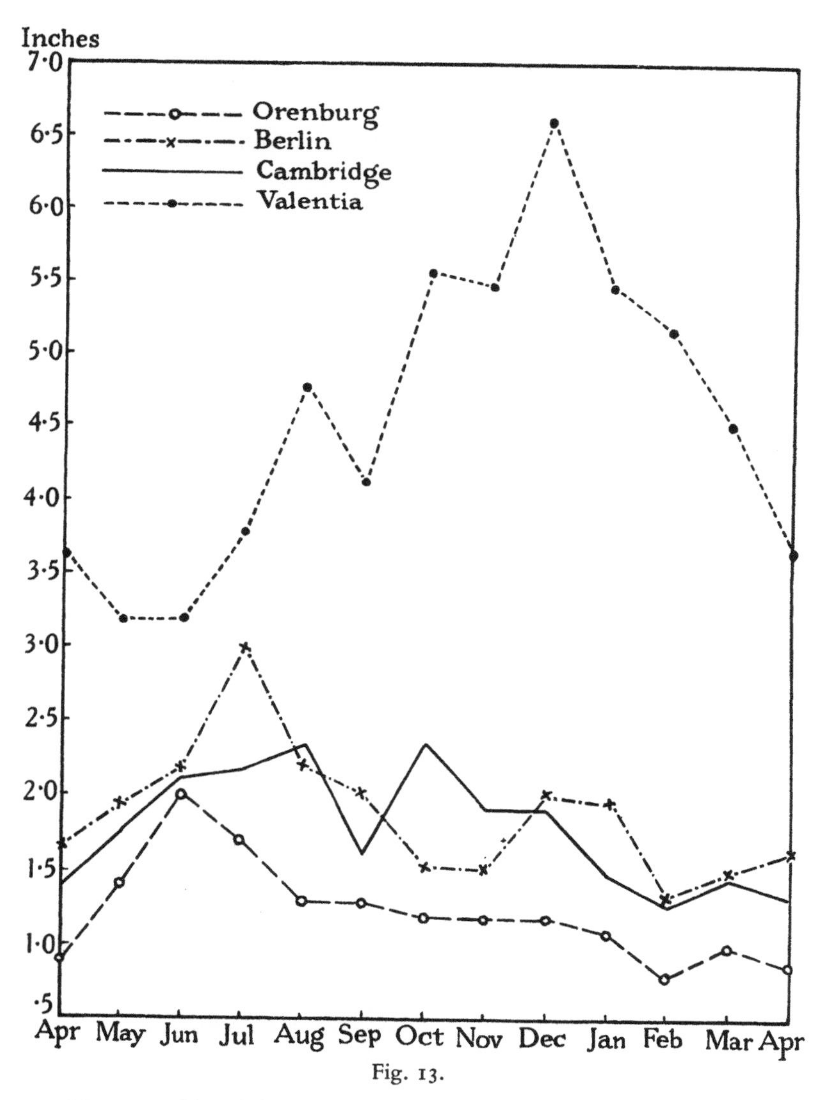


Fig. 13.

Mean monthly rainfall at Valentia, Cambridge, Berlin and Orenburg.

spring or early summer. Like the curves for the continental stations, the curve for Cambridge shows a rapid rise from April to June, but the rate of increase is not maintained, and a further slow rise leads to a maximum in August (or July–August at some other stations in the County). Thereafter, there is a fall in September followed by the assertion of a stronger oceanic influence giving a second maximum in October. This oceanic influence extends as far east as Berlin where it causes a subsidiary maximum in December which does not appear in the station farther east. After October, in Cambridge, there is a general fall to a minimum in February, the month with least rainfall in all the stations except Valentia.

## RELATIVE HUMIDITY

Cambridge is situated in one of the lowest rainfall areas in this country and the combination of a high summer temperature with a rainfall that is low must be critical for many species of plants unless there are com-

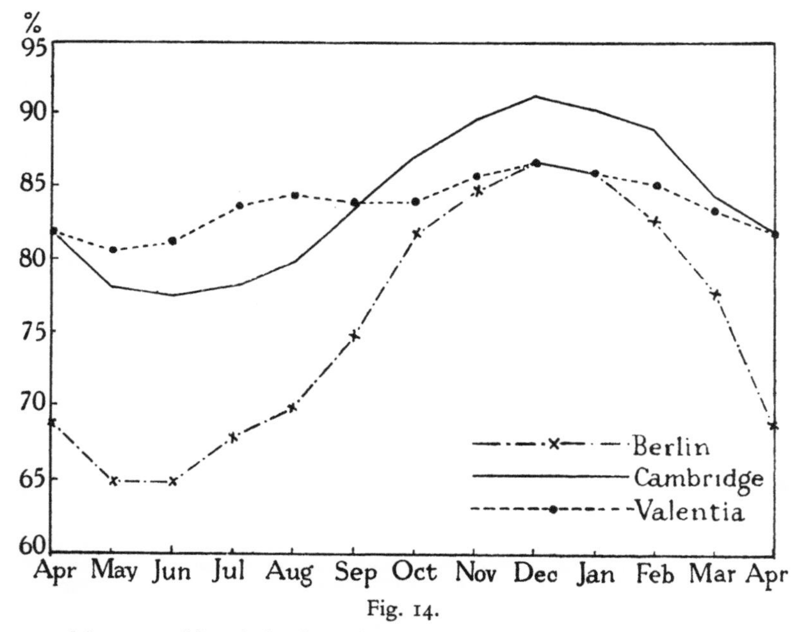


Fig. 14.

Mean monthly relative humidity at Valentia, Cambridge and Berlin.

pensating advantages. One of these is the relatively high humidity. The data are not strictly comparable; the Valentia data are for the period 1886–1910, the Cambridge data for 1924–34 (see Table 6; Fig. 14). The sustained high monthly values for Valentia are in keeping with its oceanic

Table 5. *Number of days with rain (over 0·01 in.) in each calendar month*

| | Jan. | Feb. | Mar. | Apr. | May | June | July | Aug. | Sept. | Oct. | Nov. | Dec. | Year |
|---|---|---|---|---|---|---|---|---|---|---|---|---|---|
| Valentia | 24 | 21 | 21 | 19 | 18 | 17 | 21 | 22 | 18 | 22 | 23 | 26 | 252 |
| Cambridge | 15 | 13 | 14 | 13 | 13 | 12 | 13 | 14 | 11 | 15 | 14 | 16 | 163 |
| Berlin | 15 | 15 | 15 | 13 | 13 | 13 | 15 | 14 | 13 | 14 | 14 | 15 | 169 |

Table 6. *Relative Humidity*

| | Jan. | Feb. | Mar. | Apr. | May | June | July | Aug. | Sept. | Oct. | Nov. | Dec. | Year |
|---|---|---|---|---|---|---|---|---|---|---|---|---|---|
| Valentia (1886–1910) | 86·3 | 85·5 | 83·7 | 82·0 | 81·0 | 81·5 | 83·7 | 84·6 | 84·3 | 84·3 | 86·0 | 87·2 | 84·2 |
| Cambridge (1924–34) | 90·6 | 89·2 | 84·7 | 82·1 | 78·2 | 77·7 | 78·3 | 79·8 | 83·6 | 87·4 | 90·0 | 91·4 | 84·4 |
| Berlin | 86·0 | 83·0 | 78·0 | 69·0 | 65·0 | 65·0 | 68·0 | 70·0 | 75·0 | 82·0 | 85·0 | 87·0 | 76·1 |

climate, but the average annual value for Cambridge is much the same, and the range of monthly means, although double that for Valentia, shows both a higher maximum and a lower minimum: but even this minimum is high. Berlin, on the other hand, shows a much wider range, with a maximum in December and a minimum in May–June. The depressing, enervating effect of the Cambridge climate is doubtless due to its high humidity, and even Indians complain of the heat during a hot spell in the summer. It seems likely, however, that the high humidity is due in part to local causes, but chiefly to the "continentality" of the neighbourhood. This arises from the large diurnal range of temperature with values at night and early morning falling to dew point. The result is that the readings at 9 h. and 21 h. tend to be high, particularly in later autumn and winter.

## SUNSHINE AND CLOUDINESS

The variations for bright sunshine (Table 7; Fig. 15) show relatively small differences from October to April for Valentia, Cambridge, and Berlin. From May to September, however, the differences are appreciable; Cambridge then occupies an intermediate position. Although the total number of hours of sunshine is less than that enjoyed by parts of the south coast, yet the Cambridge neighbourhood is sunny compared with the west.

The data for cloudiness (Table 8) show the nearer approach of Cambridge to Berlin during April to September: from September to April, Cambridge has the least cloudiness of all these places and also the lowest average for the year.

The preceding data show that while the climate of Cambridge is not continental, it has a number of continental features, the fuller expression of which is checked by a high humidity.

## GENERAL CONSIDERATIONS

The weather varies much from year to year and from place to place. Even minor differences in topography are significant to plant life and to man: a few feet may raise them above an accumulation of cold air or of fog. But the major variations are determined by major causes—namely, the position of Cambridge in relation to the centres of low- and high-pressure systems in north-western Europe. In general, the air moves from south-west to north-east, due to the frequency of cyclones centred (1) to the north-west of the country, or (2) directly over the British Isles or over the Channel; the former bring "orographic" rain to the west, the latter "cyclonic" rain to all parts of the country. The rainfall of Cambridge is

Table 7. *Number of hours of bright sunshine daily for each calendar month*

| | Jan. | Feb. | Mar. | Apr. | May | June | July | Aug. | Sept. | Oct. | Nov. | Dec. | Year |
|---|---|---|---|---|---|---|---|---|---|---|---|---|---|
| Valentia (1906–35) | 1·4 | 2·3 | 3·7 | 5·4 | 5·9 | 5·8 | 5·1 | 4·8 | 4·2 | 2·9 | 2·1 | 1·3 | 3·7 |
| Cambridge (1906–35) | 1·7 | 2·5 | 3·9 | 4·9 | 6·4 | 6·8 | 6·2 | 6·0 | 4·9 | 3·5 | 2·0 | 1·3 | 4·2 |
| Berlin | 1·3 | 2·2 | 3·3 | 5·6 | 7·4 | 8·2 | 7·4 | 6·9 | 4·8 | 3·1 | 1·7 | 1·1 | 4·4 |

Table 8. *Number of hours of cloudiness daily for each calendar month*

| | Jan. | Feb. | Mar. | Apr. | May | June | July | Aug. | Sept. | Oct. | Nov. | Dec. | Year |
|---|---|---|---|---|---|---|---|---|---|---|---|---|---|
| Valentia | 7·8 | 7·7 | 7·0 | 6·8 | 7·2 | 7·4 | 7·5 | 7·5 | 7·2 | 7·3 | 7·2 | 7·8 | 7·4 |
| Cambridge | 6·6 | 6·6 | 6·0 | 5·9 | 5·9 | 5·9 | 6·3 | 6·0 | 5·3 | 5·7 | 6·2 | 6·4 | 6·1 |
| Berlin | 7·3 | 7·2 | 6·6 | 6·0 | 5·6 | 5·6 | 6·1 | 5·8 | 5·5 | 6·5 | 7·3 | 7·7 | 6·4 |

essentially cycionic. At certain times of the year, particularly in spring and early summer, bitterly cold weather may be experienced: the centres of high-pressure systems are situated to the north-west and north, and cold north and north-east winds blow over East Anglia as, for example, during the destructive frost of May 1935. These winds may bring rain to our eastern shores when the west is dry. Occasionally, too, cold winds

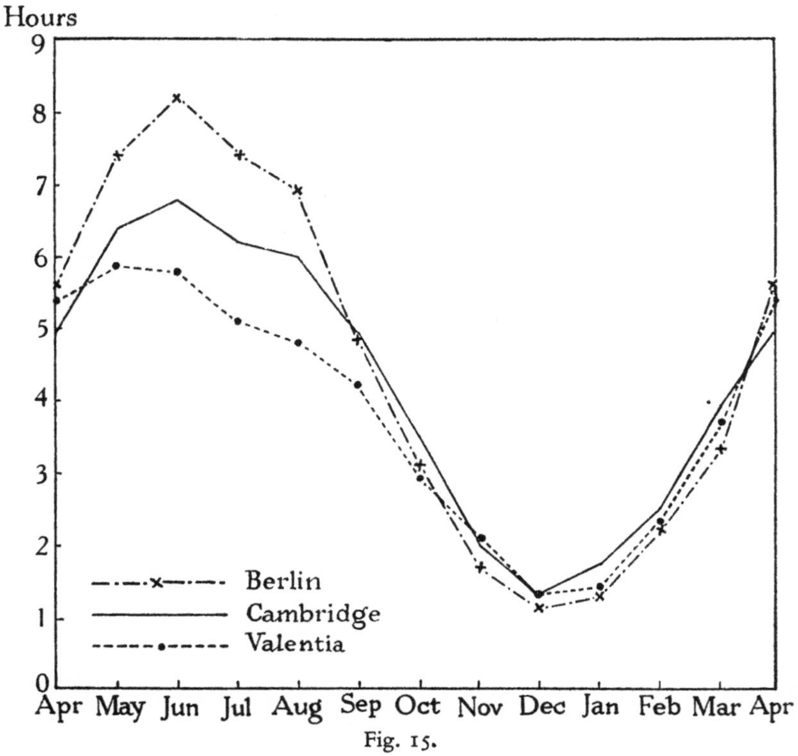


Fig. 15.

Average number of hours of bright sunshine per day at Valentia, Cambridge and Berlin.

blow from the south-east. Since, in general, these cold winds prevail during the spring, and the westerlies during the autumn, spring is colder than the autumn. At different times of the year Cambridge may experience anticyclonic conditions, but whether the weather is warm or cold depends on the origin of the anticyclone and on its position. If it is an extension of the high-pressure belt of latitude 30–35° N., then the air is warm, and long spells of dry warm weather may be enjoyed; if it is an offshoot of a winter continental anticyclone, the weather may be very cold.

These data, or the conditions they represent, are significant for plant life and agricultural practice. The rainfall, as we have seen, is low. The results of deficiency and drought would be more apparent than they are, but for the high water table of much of the neighbourhood and for the prevalence of soils with a high capacity to hold water. Where, as in Breckland, the soil is porous and non-retentive the rainfall deficiency is accentuated.[1] Thus because of the dry conditions, plants of the oceanic west may fail or grow poorly. Among the planted conifers in and around Cambridge only the xerophytic pines do at all well: spruces, hemlocks, silver firs do badly.

On the other hand, flooding may occur, and the floods of the early months of 1937 are still fresh in the memory. They were due to an exceptional run of five very wet months, January to May, when the water draining into the low Fenland depression found its normal flow to the sea checked by a combination of wind and tide.[2]

Again, although the neighbourhood of Cambridge is an area of intensive cultivation, the low temperatures of winter preclude competition with the milder south-west in the production of spring flowers. The extension into spring of cold wintry conditions and the frequency of cold winds and recurrent spring frosts until the beginning of June mean the crippling of frost-sensitive species of plants and in agricultural practice inability to grow early potatoes. Fruit-tree blossom, too, is often severely damaged. On the other hand, the greater number of hours of sunshine allows a proper maturation of wheat, and the relatively high number of hours of sunshine during September doubtless contributes to the percentage of sugar in sugar beet. For native plants, in particular for the "continental" element in our flora, sunshine is critical for the ripening of seed.[3]

[1] See p. 208 below. [2] See p. 193 below.

[3] See A. S. Watt, "Studies in the Ecology of Breckland. I. Climate, Soil and Vegetation", *Jour. Ecol.* xxiv, 117 (1936).

CHAPTER FOUR

# THE BOTANY OF CAMBRIDGESHIRE

By H. Godwin, M.A., PH.D.

THE FOLLOWING ACCOUNT OF THE BOTANY OF CAMBRIDGESHIRE is written primarily from the standpoint of the ecologist, showing the vegetation of the area in relation to geology, topography, and climate.[1] The chief vegetation types are considered in turn according to the geological formations on which they occur. As this part of England is particularly heavily cultivated, stress is specially laid on those fragmentary communities still present in a fairly natural state, such as the woodlands on Boulder Clay, the Chalk grassland, and the sedge or scrub in the undrained parts of the Fenland. Since, moreover, no other county contains so much of the old peat fens, the account deals at length with what remains of fen vegetation upon their surface.[2] Although a portion of the Breckland, with its typical soil and vegetation, comes just within the County boundary west of the River Kennett, it will not be considered here as it has been dealt with separately by Dr Watt.[3]

## THE FENLAND

The English Fenland within historic times stretched over the greater part of the area to the west and south of the Wash, extending as far north as Lincoln and as far south as Huntingdon and Cambridge (see Fig. 47). On the seaward side, the surface deposits are semi-marine silt, laid down, and afterwards occupied, during the Romano-British period.[4] On the landward side, the upper layers are peat. This peat was produced by discharge of the floodwaters of the Rivers Witham, Welland, Nene, and Ouse into the extensive shallow basin of the fens. This water entering the fens has a high mineral content, particularly that from the tributaries of the Ouse, which drain the chalk escarpment to the east. The fen peats are therefore alkaline in reaction, and support a vegetation of true "fen" type—the "Niedermoor" of German botanists. Such fens are dominated by grass-like monocotyledons of the Gramineae, Cyperaceae, and Juncaceae, and, in their drier stages, by shrubs and trees such as willow, alder, and birch.

[1] I am indebted to the Editor of the *Victoria County History* (Mr L. F. Salzman) for permission to use material prepared for a more extensive account of the vegetation of the County.

[2] See also p. 17 above for a summary of the vegetational history revealed by investigation of the successive layers of peat deposit in the Fenland.

[3] See p. 221 below.

[4] See p. 20 above, and p. 92 below.

Extremely little trace remains to-day of the original vegetation of the peat fen. Almost the whole area has been drained and brought under cultivation: its character can be recognised only by the black peaty soil, the uniform flatness, and by the deep ditches full of reeds (*Phragmites communis*) that separate fields of potatoes, cereals, and sugar beet. The continuous hawthorn hedges of the neighbouring land thin out abruptly at the fen border. Rows of planted willows, and scattered clumps of shelter trees, or small coverts, remain the only woody plants on the cultivated fen.

## I. WICKEN FEN

One of the largest and best known areas of fen still uncultivated is Wicken Fen, covering about one square mile, now in the hands of the National Trust, and lying about 10 miles to the north-east of Cambridge on the very margin of the Fenland.

The lodes, or main drainage channels traversing the area, converge at Upware, and there they communicate through sluicegates with the River Cam. The surrounding cultivated land has an entirely separate drainage system at a much lower level. The water in Wicken Fen itself is conserved in summer by the Upware sluices, and, in winter, excess water is run off whenever possible. Thus the water level in the fen changes comparatively little through the year, although the fen surface is about +7 ft. O.D. and the surrounding land has become much lower through peat wastage following draining.[1]

Not only are the soil and water-level relations in the fens thus altered from natural conditions, but peat-cutting has removed much of the fen surface, and the fen vegetation has suffered a traditional system of crop-taking maintained to some extent at the present day. Other human activities also in less degree affect the vegetational cover: these include the cutting and clearing of steep-sided lodes and drains, the consolidation and mowing of "droves" and walks, the felling of old fen scrub, and, to a small extent, propagation of rare species of plants.

1. *The Primary Succession*. In accordance with the accepted laws of the succession of plant communities, shallow open water should progress by natural accumulation of peat to shallower conditions, to a soil surface first at water level and then above it, steadily moving towards a final stable community, the climax. This succession is known as the primary hydrosere, and the climax in East Anglia may be supposed to be deciduous woodland. Human activities have obscured the original simple hydrosere relations at Wicken, but it seems clear that it probably differed little from

[1] See p. 186 below.

that of parts of the Norfolk Broads. This primary succession may be summarised as follows:

(*a*) *Aquatics*. There are no large open areas of water in the fen, and nowhere can we see zonations indicating all the early stages of the hydrosere, like those described by Miss Pallis on the Norfolk Broads. We may indeed judge the lode and ditch flora in the light of the work of Miss Pallis, and find, in appropriate depths of water, various aquatic plants, which are more or less widespread in the rest of England. The abundant forms include: *Chara*, *Elodea canadensis*, *Myriophyllum spicatum*, *Hippuris vulgaris*, *Potamogeton lucens*, *P. pectinatus*, *P. perfoliatus*, *P. crispus*, *P. densus*, *Scirpus acicularis*, *Sparganium simplex*, *Oenanthe phellandrium*, *Hottonia palustris*, *Castalia alba*, *Nymphaea lutea*, *Sparganium natans*, *Sagittaria sagittifolia*, *Butomus umbellatus*, *Alisma plantago*, *A. ranunculoides*, *Ranunculus lingua*, *Polygonum amphibium*, etc.

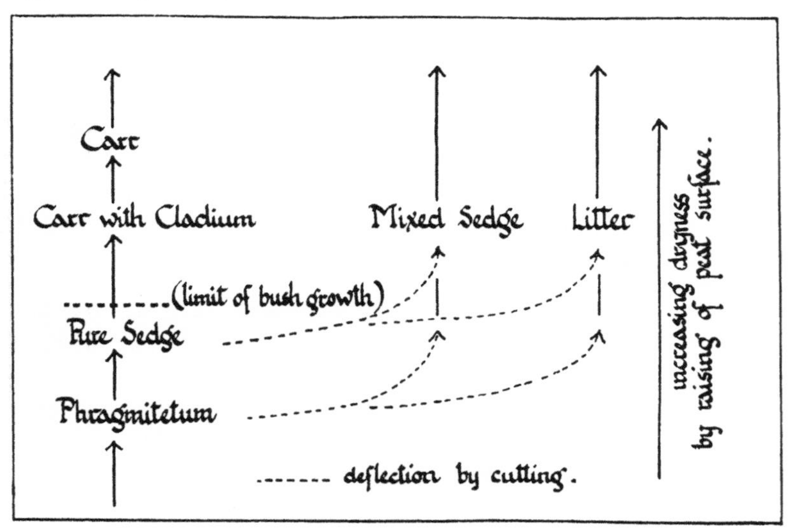


Fig. 16.

Vegetation Successions at Wicken Fen.

From H. Godwin and F. R. Bharucha, "Studies in the Ecology of Wicken Fen", *Journ. Ecology*, xx, 185 (1932).

(*b*) *Reed-Swamp* (*Phragmitetum*). The borders of lodes and ponds and the shallower drains throughout the fen show a reed-swamp of *Phragmites communis* well developed, but only at the eastern end, where the fen peat abuts on the gault clay, do *Scirpus lacustris*, *Typha angustifolia*, *Glyceria aquatica* or *Phalaris arundinacea* occupy similar marginal positions by the water side.

(*c*) *Pure Sedge* (*Cladietum*). Throughout the hydrarch succession we may reckon that development follows the gradual raising of the peat level up to and above the water level, so that we can utilise the difference between the two levels as a criterion of successional phase. It then becomes apparent that the artificially steepened banks of lodes and drains, and their artificially raised edges, give no opportunity for the successional stage next following reed-swamp to become easily evident as a zonal community. Drains and trenches, made by removing peat, are now often choked up to form habitats only little drier than those of the reed-swamp: such places support the community we have called "pure sedge". This is a closed community dominated very completely by the prickly sedge, *Cladium mariscus*. The sedge leaves grow up to 3 m. long, bending over horizontally at about 1·5 m. The luxuriance, the evergreen habit, and the thick "mattress" of dead leaves deposited between the growing shoots of *Cladium*, prevent the growth of all other plants save infrequent individuals of *Phragmites*, *Lysimachia vulgaris*, and *Salix repens* var. *fusca*.

Recent studies on the autecology of *Cladium mariscus* at Wicken have shown many features of great interest in the plant itself and the conditions under which it grows. The meristem of the plant lies below ground at the apex of a vertical stock: it is very frost-sensitive, but the wet fen peat has such a small temperature diffusivity constant, that frost seldom or never penetrates deep enough to be harmful. The meristems also offer a barrier to gas diffusion from the growing leaves to the stock and roots, but aeration of these organs takes place through the bases of dead or mature leaves. The drained upper layers of peat show strong seasonal drift in composition of the soil atmosphere determined by changing soil temperature. Values of 5 to 6 per cent of carbon dioxide at only 20 cm. depth are reached in summer with correspondingly lowered oxygen values. The soil water below the water table is apparently devoid of dissolved oxygen.

(*d*) *Bush Colonisation*. As the peat level is gradually raised, the pure sedge is invaded by bushes, which are dispersed with great rapidity by seeds. The most abundant species is *Rhamnus frangula* (*Frangula alnus*); next come *Rhamnus catharticus*, *Salix cinerea* and *Viburnum opulus*, and lastly there is a very small percentage of *Crataegus monogyna*, *Prunus spinosa*, *Ligustrum vulgare* and *Myrica gale*. Both dissemination and establishment are irregular, and early stages of bush colonisation have a very heterogeneous structure, which disappears, however, as the sedge patches are invaded marginally or over their whole extent, and the bushes come to form a complete and uniform cover over the whole area. Birds are probably the most important agents of dispersal of the *Rhamnus* bushes, especially the large flocks of migratory fieldfares which visit the fen in autumn as the drupes ripen. On the other hand, a very large part of the fruit crop falls to the ground, and, as the stones are exposed by the drying of the fruit, they are taken by fieldmice, which often gather them into stores, where, if forgotten or abandoned, they may germinate.

(*e*) *Carr* (*Franguletum*). As the canopy of the young buckthorn becomes closer there follows, especially where colonisation has been dense, a well-marked phase in which *Cladium* and associated species of the previous stage are killed out; their dead leaves still hang in the crotches of the branches when no living plants persist. The ground becomes almost bare beneath the bushes and a characteristic shade-tolerant flora enters. Thi usually includes the marsh ferns, *Dryopteris thelypteris*, *Agrostis stolonifera*, *Urtica dioica*, and non-flowering shoots of *Lysimachia vulgaris*, *Symphytum officinale* and *Iris pseudacorus*. *Convolvulus sepium* is often a conspicuous feature of early phases of scrub formation, twining round the "drawn-up" reed stems, and the sallow, *Salix cinerea*, is a typical pioneer shrub, succumbing early to the competition of other bushes.

As the carr ages, the bush density diminishes and it seems certain that dominance passes from *Rhamnus frangula* to *R. catharticus*. The mechanism of this displacement is uncertain, but it probably involves an extensive "die-back" disease caused by the fungi *Nectria cinnabarina* and *Fusarium* sp. To this, the older *R. frangula* bushes seem very susceptible; the fungi gain entrance by snags and rapidly kill the bushes. *R. catharticus*, which is not so attacked, increases greatly in the later stages of carr development. At this stage, also, the bushes take on a tree form with central trunk and branches limited to a close crown—a strong contrast to the earlier scrub in which each bush stool has many trunks, and branching is extremely diffuse and extensive. In later stages of development, *Viburnum opulus* may be of importance: it straggles extensively under the shade of the other bushes, rooting at the places where shoots touch the soil, to form quite impenetrable tangles.

There is no deciduous fen woodland on Wicken Fen, but in many places birch (*Betula alba*) is spreading from seed; there are also a few good-sized oaks (*Quercus robur*), a colony of grey poplar (*Populus canescens*), and many scattered ashes. It is a remarkable fact that, although the fen peat contains pollen and wood of alder (*Alnus glutinosa*) in great quantity, the few planted trees now growing on the fen are not able to spread, although they produce abundant viable seed.

2. *Deflected Successions.* Over the greater part of the fen the natural succession already described does not take place, for the fen vegetation is cut at intervals, as a rule, of either one year or four years, and the crop is used for litter (cattle-bedding) or for thatching. On these lines, cutting

has probably gone on for centuries. It is evident that although bush colonisation will be prevented by this practice, yet the peat level will continue to rise, and new successions, still subject to the cutting factor, will take place. These will, however, differ from the primary succession and give rise to communities not found therein. Such are the "mixed sedge" and "litter" which cover almost 50 per cent of the fen. These two communities occur on peat often much higher above water level than even mature carr, and as soon as the "deflecting factor" of cutting is removed, they very rapidly become colonised by bushes and give rise to carr. It is quite likely that the carr development stages we have already mentioned as part of the prisere may more properly be referred to these shortened successions.

(*f*) *Mixed sedge* (*Cladio-Molinietum*). Vegetation in which *Cladium mariscus* and *Molinia cœrulea* are more or less co-dominant has long been cut by fenmen for thatch or for fuel. In it, the sedge (*Cladium*) is rather less vigorous than in the pure sedge, a reflection of the influence of the drier habitat upon its summer growth rates. It is scattered through with *Phragmites*; while flowering plants such as *Eupatorium cannabinum*, *Angelica sylvestris*, *Peucedanum palustre*, *Lysimachia vulgaris*, *Hydrocotyle vulgaris* and *Salix repens* var. *fusca*, occur sparsely throughout, though they are much more conspicuous in the season following a sedge crop than afterwards.

It is clear that this vegetation has arisen by persistent cropping of the vegetation, which serves to exclude bushes, but which allows continued peat growth and the ingress of plants characteristic of drier habitats but not susceptible to cutting. *Molinia* is, of course, the most important of these, for it produces annual photosynthetic shoots, and its tuberised stem bases are unharmed by winter scything: in this, it contrasts greatly with *Cladium*, the leaves of which are evergreen, and the loss of which seriously damages the plant. Bush invasion, especially by *Rhamnus frangula*, is very rapid when cutting is discontinued.

(*g*) *Litter* (*Molinietum*). When fen owners have cut the vegetation at short intervals, such as one year, the sedge has been rapidly killed out. The Molinietum, thus formed, like the Cladio-Molinietum, contains much *Phragmites*. In it, however, *Carex panicea* and *Juncus obtusiflorus* are subdominant, and several smaller plants, encouraged by the removal of taller dominants, also appear. These include *Succisa pratensis*, *Thalictrum flavum*, *Cirsium anglicum*, *Valeriana dioica*, and *Orchis incarnata*. The yearly crops of litter are used for cattle-bedding or for coarse chaff; the straight boundaries of the litter communities coincide with the limits of different owners' plots, and betray their origin.

It has been demonstrated experimentally that a mixed sedge community can be altered, in ten years by annual cutting, to something that fairly closely resembles litter. Conversely, cessation of cutting will cause the disappearance of such species as *Carex panicea* from late stages of Cladio-Molinietum.

There are, in fact, any number of inter-grades between the communities we now describe, and they lie upon a series of deflected successions suffering cutting of different intensities; when left uncut, they rapidly form carr.

The droves in the fen, which are cut twice or thrice a year, might be considered to show successions still further deflected from the prisere. Certainly, the main drove, which has been in existence for at least two centuries, has a most characteristic flora, very rich in species.

From time to time, areas of mature carr are cleared of bushes, and very striking vegetational changes follow. Innumerable seedlings of a great number of species rapidly establish themselves, and among them are shoots of *Iris*, *Rubus caesius*, *Dryopteris thelypteris* and other relicts of the previous shade phase. The very rich herbaceous vegetation, if undisturbed, quickly reverts to carr.

## II. Other Remnants of Fen Vegetation

Around the fen margins there still exist a few much-modified areas of fen vegetation. After Wicken, the largest of these is Chippenham Fen, which lies 4 to 5 miles north of Newmarket, where the Breckland sands come down to the fens. Over most of this fen, peat-cutting has left deep trenches, now filled densely with *Cladium mariscus*. Associated with it, but especially on the ridges, *Molinia* is abundant, while the following species are more or less frequent: *Schoenus nigricans*, *Juncus obtusiflorus*, *Angelica silvestris*, *Eupatorium cannabinum*, *Lythrum salicaria*, *Urtica dioica* *Valeriana officinalis*, *Serratula tinctoria*, and *Scrophularia nodosa*. *Phragmites* is abundant throughout, and there is close general resemblance to the Cladio-Molinietum at Wicken, though bush colonisation is much sparser. Finally, Chippenham Fen is still the home of *Pinguicula vulgaris*, *Aquilegia vulgaris*, *Selinum carvifolium* and *Carex pulicaris*.

Other remnants of fen vegetation occur in the Cam Valley at Dernford Fen, now rapidly drying up, at Quy Waters and Quy Fen. The peat is generally very shallow, and the fen species persist precariously in ditches and pools.

The County flora also includes species such as *Stratiotes aloides*, *Teucrium scordium*, *Villarsia nymphaeoides* hanging on in these sites: while at Wicken still persist *Liparis loeselii*, *Ranunculus lingua*, *Peucedanum palustre*,

*Lathyrus palustris* and *Myrica gale*. *Viola stagnina* has not been seen for some years, and it has probably followed *Senecio paludosus*, *S. palustris*, *Sonchus palustris* and *Cicuta virosa* into the list of species now extinct. It is possible that *Typha minima* grew quite recently at Wicken. Throughout the County the fen lodes and their margins naturally still carry an abundant selection of the old fen species, although few of these are rare.

Somewhat different in character from these true relics of the fens, are a few sites on the Gault or Chalk, where local conditions formerly led to the growth of small fens or even of acidic bogs. These sites have now been drained, but from Hinton, Teversham, and Sawston Moors, the following have been recorded: *Drosera rotundifolia*, *D. anglica*, *D. intermedia*, *Pinguicula vulgaris*, *Malaxis paludosa*, *Scirpus caespitosus*, *Eriophorum angustifolium*, *Carex dioica*, *Molinia cœrulea*, *Sphagnum cymbifolium* and *Splachnum ampullaceum*. Triplow Holes is a site on the chalk where fen species still survive, and where *Cladium* has a local dominance.

### III. The Silt Fens

The silt which forms the fen soil in the northern part of the County round Wisbech (see Fig. 29) has probably been cultivated, apart perhaps from the Saxon period, ever since its deposition in Romano-British times.[1] Though "natural" vegetation is absent from it, it would be of great interest to work out the progress of invasion and establishment of species in the area during the quite definite period since its origin in brackish water. So far this has not been done.

At Foulanchor, near Wisbech, an area of reclaimed salt marsh falling within the County boundary brings a number of maritime species into the County flora. The tidal influence which formerly extended far inland up the fen rivers has no doubt been responsible for inland records of the more tolerant maritime species, such as *Scirpus maritimus* at Littleport, Sutton and Upware, and *S. tabernaemontani* at Littleport.

## THE LOWER GREENSAND AREA

The outcrop of Lower Greensand in the County is not extensive. Some of it supports much of the market gardening and orchard area near Cambridge, and the fields show typical psammophilous weeds. Only at Gamlingay, in the extreme south-west, does heath develop on it, and even here extensive tree-planting, felling and pig-keeping have greatly altered the natural vegetation. Of the former heath dominants *Calluna vulgaris* and *Deschampsia flexuosa* are still abundant. The following also occur: *Teesdalia nudicaulis*, *Galium saxatile*, *Ulex europaeus*, *Cytisus scoparius*, *Luzula multi-*

[1] See p. 92 below.

*flora, Nardus stricta, Aira praecox, Anthoxanthum odoratum.* Equally typical of heath conditions are the mosses, *Polytrichum piliferum, P. juniperinum, Bryum roseum, Dicranum scoparium* var. *orthophyllum, Hypnum schreberi, Brachythecium albicans* and *B. purum.* Until recently *Tilia cordata* and *Quercus sessiliflora* apparently grew naturally here.[1]

The Greensand outcrop at Gamlingay is also noteworthy because until 1855 it carried a large acidic peat bog yielding species quite characteristic of the surfaces of raised bogs. The moors at Hinton, Teversham, and Sawston, already mentioned, may have been more or less similar, but on a large scale they are rare in the east of England. Peat investigations in the Woodwalton and Yaxley areas of fen, south of Peterborough, show that the peat fens locally passed into the condition of raised bogs, and that limited marginal areas of fen surface retained this character and typical flora until quite recent times. Most of the raised bog species from Gamlingay are now extinct, but the records and herbarium specimens are sufficiently convincing. They include the following:

*Hypericum elodes*
*Drosera rotundifolia*
*Oxycoccus quadripetalus*
*Littorella lacustris*
*Malaxis paludosa*
*Narthecium ossifragum*
*Potamogeton polygonifolius*
*Scirpus pauciflorus*
*Eriophorum angustifolium*
*Rynchospora alba*
*Carex dioica*
*C. stellulata*
*Lycopodium clavatum*
*L. inundatum*
*Sphagnum angustifolium*
*S. cuspidatum*
*Archidium alternifolium*
*Splachnum ampullaceum*
*Aulacomnium palustre*
*Odontoschisma sphagni*
*Aneura pinguis*
*Hypnum revolvens*
*H. stellatum*
*Philonotis fontana*
*Polytrichum commune*

## THE BOULDER CLAY

### I. Woodland

The Boulder Clay in Cambridgeshire lies in two large patches, on the eastern and western sides of the County (see Fig. 29). Place-name evidence shows that these were the wooded districts of the County in the Anglo-Saxon period.[2] In Domesday times also, the distribution of woodland varied sympathetically with that of the clay (see Fig. 17). To-day, by far the greater part of the sparse woodland of the County is to be found in these two areas. These woods are indeed almost the only semi-natural vegetation

[1] Boulder Clay overlying Greensand in some places, and Gault in others, also occurs at Gamlingay, and supports woodland. See p. 54 below.
[2] See p. 103 below.

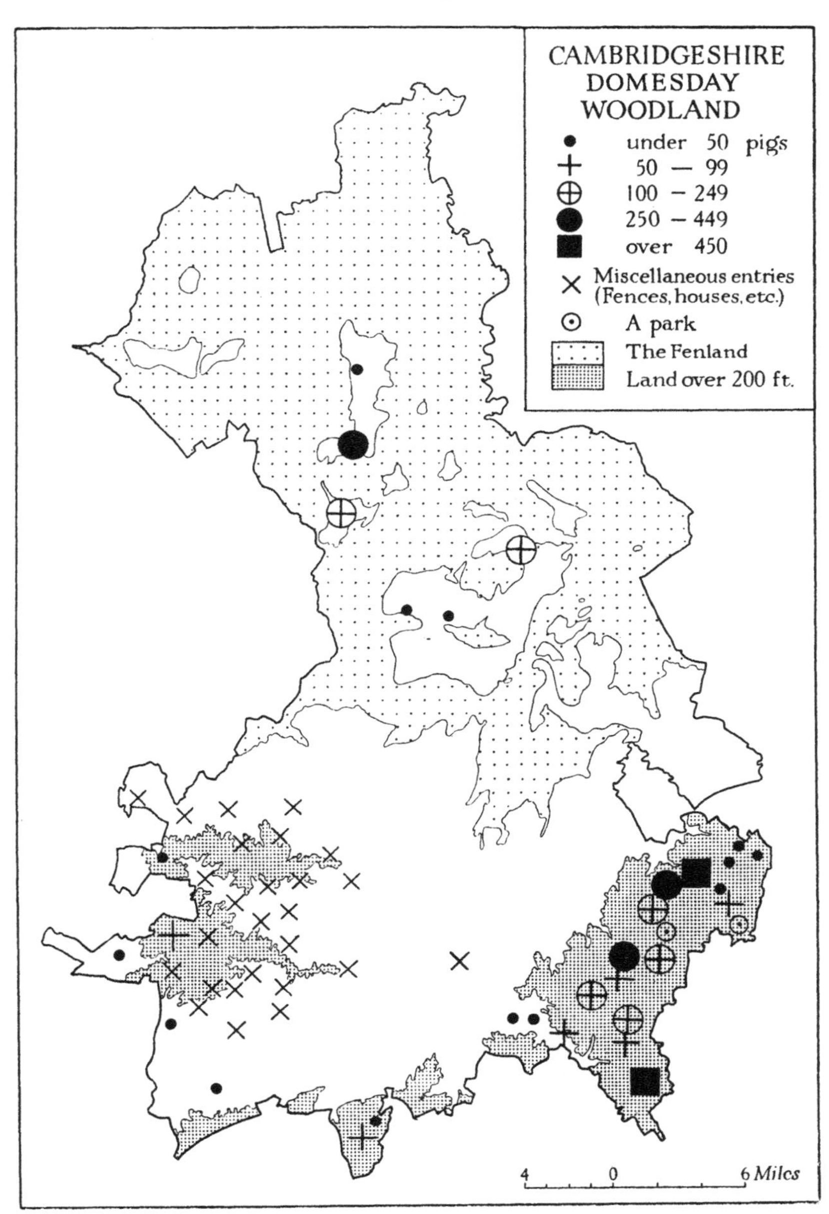


Fig. 17.

From H. C. Darby, "The Domesday Geography of Cambridgeshire", *Proc. Camb. Antiq. Soc.* xxxvi, 49 (1936).

on the Boulder Clay, and merit special consideration. They are almost all of the type known as "coppice with standards", in which standard oak trees (*Quercus robur*) project from a dense shrub layer consisting mainly of hazel (*Corylus avellana*), which is coppiced at rather irregular intervals. They were classified by Adamson many years ago as of the (ash)-oak-hazel type, derived by exploitation from the natural woods by suppression of the ash, which, however, by its strong regenerative powers still gives clear evidence of its natural status. Affinities with the ash woods of calcareous soils are shown by the frequency of calcicole shrubs such as the spindle tree, *Euonymus europaeus* and the wayfaring tree, *Viburnum lantana*, and by herbs in the undergrowth such as *Mercurialis perennis*, *Viola silvestris*, and *Hypericum hirsutum*. The bush species are numerous and include *Acer campestre*, the maple, which is often coppiced with the hazel; both species of hawthorn; the *Crataegus monogyna*, much less frequent as shrub undergrowth than *C. oxyacanthoides*; privet (*Ligustrum vulgare*); dogwood (*Cornus sanguinea*); and blackthorn (*Prunus spinosa*). The maple, privet and dogwood are more frequent than in pure *Quercetum roburis*. Some woods also contain *Viburnum opulus*, *Salix caprea*, *S. cinerea*, *Prunus cerasifera* and *Daphne laureola*.

The early work of Adamson on Gamlingay Wood shows most clearly the dependence of woodland characters upon soil. Most of the wood has a calcareous marl soil where the Boulder Clay overlies Gault, but there is a smaller inland area with a loam soil where the Boulder Clay is above Greensand. The two regions differed strikingly from one another. Abundant coppiced species on the calcareous clay were the ash, maple and hazel, but these were infrequent on the non-calcareous loam. Conversely, the two birches (*Betula alba* and *B. pubescens*) were frequent on the loam but absent from the clay. Similar wide divergences were recognisable between the undergrowth communities. Adamson recognised on the clay soil the four following societies:

(1) *Filipendula ulmaria* society—high summer water content and low light intensity.

(2) *Filipendula ulmaria-Deschampsia caespitosa* society—with high water content but lighter than (1).

(3) *Mercurialis perennis* society—on drier soils and with a wide light range.

(4) *Fragaria vesca* society—in conditions intermediate between those of (2) and (3).

On the loam soil, he recognised two societies, a *Pteridium aquilinum-Holcus mollis* society on the heavier loam, and a *Holcus mollis* society on the sandier loam.

Through the Boulder Clay woods of the rest of the County, Adamson recognised the same communities, the drier, such as the *Mercurialis* society, especially in soils over the Chalk, and the wetter, such as the *Filipendula* society, in soils over the Gault. In general, however, the impermeability of all the Boulder Clay soils leads to winter water-logging which has a marked local influence on the ground flora.

A generalised idea of the ground flora can be gathered from the following lists of species in Hardwick Wood:

Dense old coppice:

*Primula elatior*
*Viola riviniana*
*V. silvestris*
*Circaea lutetiana*
*Sanicula europaea*
*Geum urbanum*
*Hedera helix*
*Ajuga reptans*
*Mercurialis perennis*
*Scilla non-scripta*
*Arum maculatum*
*Listera ovata*
*Orchis maculata*
*Neottia nidus-avis*
*Habenaria bifolia*

Recently coppiced areas show in addition:

*Anemone nemorosa*
*Ranunculus ficaria*
*Viola hirta*
*Hypericum hirsutum*
*Lathyrus silvestris*
*Filipendula ulmaria*
*Rubus caesius* et spp.
*Epilobium angustifolium*
*E. hirsutum*
*E. montanum*
*Angelica silvestris*
*Galium aparine*
*Arctium minus*
*Cirsium palustre*
*Primula vulgaris*
*Solanum dulcamara*
*Scrophularia nodosa*
*Prunella vulgaris*
*Lamium galeobdolon*
*Stachys silvatica*
*Rumex viridis*
*Tamus communis*
*Juncus effusus*
*Carex silvatica*
*C. glauca*
*Deschampsia caespitosa*
*Brachypodium silvaticum*

Further interesting species present in other woods of the same type are *Paris quadrifolia, Helleborus viridis, H. foetidus, Conopodium denudatum, Geum intermedium,* and *Melampyrum cristatum.* On the other hand, it is remarkable that *Oxalis acetosella, Adoxa moschatellina,* and *Allium ursinum,* should be extraordinarily infrequent: the foxglove (*Digitalis purpurea*) is quite absent. The true oxlip, *Primula elatior*, was shown by Miller Christie to be confined in this country to the Boulder Clay areas of East Anglia; and in Cambridgeshire this restriction is very clear,[1] but the status of the

[1] Just as the oxlip and other woodland species like *Paris quadrifolia* and *Daphne laureola* are confined to this formation, so in pastures are *Genista tinctoria* and *Trifolium ochroleucum*; in wood-margins and hedgerows, *Melampyrum cristatum*; and in cultivated fields *Linaria elatine, L. spuria*, and the very rare *Bupleurum rotundifolium* and *Euphorbia platyphyllos*.

plant in our woods and its relation to the primrose (*Primula vulgaris*) are still extremely uncertain, though the hybrids are both abundant and fertile.

The moss flora is not extensive, the commonest species being *Thuidium tamariscinum*, *Brachythecium rutabulum*, *Catharinea undulata*, *Hylocomium triquetrum*, *H. squarrosum*, *Fissidens taxifolius*, *Eurhynchium praelongum*, and *Porotrichum alopecurum*. Elm woods on the Boulder Clay, mostly *Ulmus minor*, can probably be taken as plantations, such as that at Knapwell in which a field system is still recognisable.

## II. Scrub

In the south-west of the County, the Boulder Clay cover over Gault has proved so intractable, on account of its very high clay content and deficiency in phosphate, that large areas were allowed to go out of cultivation.[1] Extensive areas of hawthorn scrub of different ages, in consequence, now occupy the ground, and it is possible to make out the main stages of a secondary succession towards woodland. Besides the dominant *Crataegus monogyna*, the young scrub shows frequent *Rosa canina* and *Prunus spinosa*; while the following are either occasional or rare: *Rosa arvensis*, *Rubus fruticosus*, *Ligustrum vulgare*, *Rosa micrantha*, *Rubus caesius*, *Acer campestre*, *Rhamnus catharticus* and *Viburnum lantana*. Along with these, scattered trees of oak or ash are found, and *Ulmus minor* often extends by suckering from nearby hedges. The early stages of bush growth show a remarkable flora of ruderal and pasture species, strongly influenced by very heavy rabbit-grazing and by the local water-logging that follows clogging of the field drains. As the bush canopy closes, this ground flora becomes sparser, and internal competition between the bushes grows, until there is produced a dense scrub of pure *Crataegus monogyna* bushes 5 to 6 m. high, and well spaced apart. The ground is practically bare, but there may be present a very few weakly plants of *Viola hirta*, *Mercurialis perennis*, *Urtica dioica*, *Brachythecium purum*, *Eurhynchium praelongum*, *Fissidens taxifolius*, *Hylocomium triquetrum* and *Mnium undulatum*. It has recently been demonstrated that this scrub shows stages in the development of a new natural soil profile from the old puddled clay surface, and, with further thinning of the old hawthorns, further entry of trees might be expected.

Throughout the succession, animal factors seem to be of great importance: rabbits, mice, woodpigeons, and magpies are present in very great density, while caterpillars wreak great havoc at times below the hawthorn canopy.

[1] See pp. 131 and 150 below.

## THE CHALK FORMATIONS

### I. Grassland

The wide stretches of Chalk grassland on the North and South Downs and on Salisbury Plain are such a uniform and well-characterised community that the Chalk grassland of Cambridgeshire cannot fail to have special interest. Comparatively little of it remains untouched by cultivation, but parts of Newmarket Heath, the Gogmagog Hills, and Royston Heath (just outside the County) are still more or less natural. The old Roman road (the *Via Devana*), the Devil's Dyke, and the Fleam Dyke (see Figs. 20 and 21), are now also clothed with grass communities and bear most of the typical and some of the rare species of chalk grassland.

A rough indication of the composition of the plant community can be gained from the following group of species collected in an area of a few hundred square yards on Royston Heath. It was obtained by a student class, and represents the results of twenty-nine random throws of a quadrat of one decimetre square. The species are listed in order of the frequency with which they occur in the twenty-nine samples; the figure after each species shows the number of quadrats in which it appeared; the letter before it shows its life-form in Raunkiaer's terms (H = Hemicryptophytes—buds in surface layers of soil; Ch = Chamaephytes—buds close above ground; G = Geophytes—buds below soil; Th = Therophytes—annuals).

| | | | | | |
|---|---|---|---|---|---|
| H | *Festuca ovina* | 28 | H | *Asperula cynanchica* | 6 |
| Ch | *Helianthemum chamaecistus* | 24 | H | *Ranunculus bulbosus* | 6 |
| H | *Poterium sanguisorba* | 20 | Ch | *Thymus serpyllum* | 5 |
| G | *Carex glauca* | 19 | Th | *Linum catharticum* | 3 |
| H | *Filipendula hexapetala* | 18 | H | *Campanula rotundifolia* | 2 |
| H | *Koeleria gracilis* | 17 | H | *Achillea millefolium* | 1 |
| H | *Plantago lanceolata* | 16 | Th | *Gentiana amarella* | 1 |
| H | *Avena pratensis* | 12 | H | *Hieracium pilosella* | 1 |
| H | *Bromus erectus* | 12 | H | *Lotus corniculatus* | 1 |
| H | *Briza media* | 10 | H | *Pimpinella saxifraga* | 1 |
| H | *Plantago media* | 10 | H | *Scabiosa columbaria* | 1 |
| H | *Hippocrepis comosa* | 9 | H | *Taraxacum officinale* | 1 |
| H | *Cnicus acaulis* | 8 | | | |

In view of the very small area examined, it is extraordinary how closely this corresponds to Chalk grassland examined by Tansley and Adamson on the South Downs: of the fifteen most constant species given by these authors, fourteen are represented above. Many other highly characteristic species are to be found in other parts of the grassland on Royston Heath; among them are the following: *Leontodon hispidus, Avena flavescens, Galium verum, Primula veris, Carlina vulgaris, Polygala vulgaris, Daucus carota*, *Anthyllis vulneraria*, *Campanula glomerata*, *Euphrasia officinalis*, *Thesium humifusum*, and *Astragalus danicus*. Two species of special interest are the Pasque flower (*Anemone pulsatilla*) and the bee orchis (*Orchis apifera*).

In the above list, the preponderance of hemicryptophytes is particularly striking; it may perhaps reflect on the one hand grazing, on the other, summer drought to which the porous chalk soil is very liable. At Royston, there is much local differentiation of grassland types from short rabbit-grazed turf on the hill crests, with *Festuca ovina* dominant, to dense thick turf dominated by *Bromus erectus*, on the deeper soils of the slopes and valley bottoms. On disturbed soils, *Arrhenatherum avenaceum* becomes prominent.

The general uniformity of the Chalk grassland can be illustrated by comparing with the Royston list those for two separate one-metre quadrats, one on the old Roman road and the other on the Fleam Dyke. Frequencies of the species are given by the conventional symbols.

| Northern end of the *Via Devana* | | Fleam Dyke[1] | |
|---|---|---|---|
| *Festuca ovina* | co–d | *Festuca ovina* | a |
| *Koeleria gracilis* | ,, | *Poterium sanguisorba* | co–d |
| *Avena pratensis* | ,, | *Hieracium pilosella* | va |
| *Scabiosa columbaria* | a | *Carex glauca* | sd |
| *Thymus serpyllum* | f to a | (In descending order of frequency) | |
| *Lotus corniculatus* | ,, | *Helianthemum chamaecistus* | |
| *Helianthemum chamaecistus* | ,, | *Briza media* | |
| *Asperula cynanchica* | ,, | *Koeleria gracilis* | |
| *Poterium sanguisorba* | ,, | *Avena pratensis* | |
| *Galium verum* | ,, | *Thymus serpyllum* | |
| *Carex glauca* | ,, | *Hippocrepis comosa* | |
| *Centaurea nigra* | ,, | *Lotus corniculatus* | |
| *Euphrasia officinalis* | f | *Leontodon hispidus* | |
| *Daucus carota* | o to f | *Asperula cynanchica* | |
| *Plantago media* | ,, | *Galium verum* | |
| *P. lanceolata* | ,, | *Cirsium acaule* | |
| *Cirsium acaule* | ,, | *Plantago media* | |
| *Anthyllis vulneraria* | ,, | *P. lanceolata* | |
| *Linum alpinum* var. *anglicum* | ,, | *Scabiosa columbaria* | |
| *Filipendula hexapetala* | ,, | *Pimpinella saxifraga* | |
| *Onobrychis sativa* | ,, | *Centaurea nigra* | |
| *Phleum pratense* | ,, | *Linum cartharticum* | |
| | | *Campanula rotundifolia* | |
| | | *Euphrasia officinalis* | |
| | | *Anthyllis vulneraria* | |
| | | *Carlina vulgaris* | |
| | | *Pinus silvestris* (one seedling) | |

[1] This list is taken from A. G. Tansley, *Types of British Vegetation* (1911), p. 178.

The moss flora of the Chalk grassland is equally characteristic; the following representative list is given by Dr P. W. Richards:

| | |
|---|---|
| *Camptothecium lutescens* } abundant | *Seligeria pauciflora* |
| *Brachythecium purum* } abundant | *Phascum curvicolle* |
| *Hypnum molluscum* } abundant | *Brachythecium globosum* |
| *H. chrysophyllum* } abundant | *Fissidens decipiens* |
| *Trichostomum flavo-virens* | *Hypnum cupressiforme* |
| *T. tortuosum* | var. *tectorum* |
| *Cylindrothecium concinnum* | var. *elatum* |
| *Ditrichum flexicaule* | *Pottia lanceolata* |
| *Weisia crispa* | *P. recta* |
| *W. microstoma* | *Tortula pusilla* |
| *Encalypta vulgaris* | *Thuidium abietinum* |

## II. Woodland

There are many indications that, where grazing allows, scrub will invade the Chalk grassland, and that the incoming bushes and trees will often form dense thickets. Hawthorn, blackthorn, dogwood, and *Rhamnus catharticus*, are usually the commonest shrub species: the evergreen yew and juniper occur but sparsely.

There is little evidence about the natural woodland vegetation of the Chalk. Though several beech woods exist, they are either plantations or have been much altered by planting, and the beech regenerates feebly in them. In the upper peats by the fen margin, however (e.g. Wicken), quite high percentages of beech pollen are to be found, which suggests that natural beech woods were recently growing nearby.

The beech woods are small and the floor is often wind-swept, so that the undergrowth is sparse. It commonly includes in the shrub layer *Ligustrum vulgare*, *Rubus fruticosus*, *Ilex aquifolium*, *Sambucus nigra*, and sometimes *Taxus baccata*. In the herb layer, there are commonly *Poa nemoralis*, *Festuca rubra*, *Brachypodium silvaticum*, *Nepeta glechoma*, *Fragaria vesca*, *Myosotis silvatica*, *Listera ovata*: less common are *Cephalanthera grandiflora*, *Orchis maculata* and *Monotropa hypopithys*.

CHAPTER FIVE

# THE ZOOLOGY OF CAMBRIDGESHIRE

## Edited by A. D. Imms, F.R.S.

(With contributions by M. D. Brindley, W. S. Bristowe, J. E. Collin, H. St J. K. Donisthorpe, J. C. F. Fryer, A. D. Imms, G. J. Kerrich, A. G. Lowndes, W. H. Thorpe, H. Watson, and H. E. Whiting)

AMONG WRITINGS ON THE ZOOLOGY OF THE COUNTY, THE manuscript catalogue of insects, and related animals, which was compiled by the Rev. L. Jenyns (afterwards Blomefield), deserves first mention. Its author lived at Bottisham in the early part of the last century and his observations were made prior to 1849. His list makes it possible to ascertain, in a general way, what species have declined or become extinct during the last century or so. The catalogue is kept in the University Museum of Zoology. When the British Association visited Cambridge in 1904, there was produced the *Handbook to the Natural History of Cambridgeshire* (edited by J. E. Marr and A. E. Shipley). This has remained the only general account of the zoology of Cambridgeshire.

Between 1923 and 1932 there appeared *The Natural History of Wicken Fen*, edited by Prof. J. Stanley Gardiner. This work makes a notable advance on previous knowledge of the zoology of the County. In 1934, there came *The Birds of Cambridgeshire*, by D. Lack. Finally, the present year will see the publication of the first volume of the *Victoria County History of Cambridgeshire* which will contain the most up-to-date and detailed account of the local fauna.[1]

### MAMMALIA

The mammals are rather poorly represented in the County. The absence of any large wooded areas is regarded as being one of the contributing causes, while the reclaimed Fenland seems to be unsuitable for supporting any considerable mammal population. Among the bats is included the scarcest mammal of the County, viz. the mouse-ear bat (*Myotis myotis*). A living specimen of this creature was recorded from Girton in 1888, and was doubtless a wanderer from the continent. The whiskered bat (*M. mystacinus*) and Natterer's bat (*M. nattereri*) are scarce, but apparently resident, species. Among other species, Daubenton's bat (*M. daubentoni*),

[1] I would like to acknowledge a general indebtedness to the Editor (Mr L. F. Salzman) for allowing us to use material prepared for the *Victoria County History of Cambridgeshire*.

the long-eared bat (*Plecotus auritus*), and the barbastelle (*Barbastella barbastellus*) also occur, but the last-named appears to be comparatively rare. The fox is general except in the Fenland, where it is a straggler. The badger has been noted occasionally in different parts of the County, and the otter occurs in the river near to, and above, Cambridge, but is rather infrequent. The stoat and weasel are plentiful, but the pine marten and true polecat are extinct. Some of the later records of the last-named species probably refer to polecat-ferrets and not to genuine wild specimens. The common shrew, the pigmy shrew, and the water shrew all occur, the first-named being the commonest. The dormouse appears to be very rare, and, of the voles, the water vole and the short-tailed vole are prevalent, while the bank vole is uncommon. The long-tailed fieldmouse appears to be local, and the harvest mouse has not often been recorded. The red squirrel occurs in various localities, while the American grey squirrel has only been occasionally reported.

## AVES[1]

In general, a county is a most unsatisfactory unit for ecological and natural history studies. This is even more obvious when dealing with birds than it is with more sedentary animals. Indeed, the only reason for choosing counties as a basis for studies of bird distribution is that they provide accurately demarcated areas of convenient size. Accordingly, the object of this brief sketch is to call attention to the main bird habitats of the Cambridge district without deference to county boundaries.[2]

To the west of Cambridge lies a countryside of heavy clay soils (see Fig. 29), mostly under cultivation, with some small mixed deciduous woods and copses in which oak predominates. Here the birds are typical, in general, of the Midlands, although the absence of larger woods with old trees restricts the fauna considerably. Indeed, as D. Lack has pointed out, Cambridge itself is almost the only part of the district where old deciduous trees are numerous, and where tree-climbing and hole-nesting species are common.[3] Five species of tits, three woodpeckers, and the stockdove, are all associated with old trees and may be observed on the Backs. Here, too, the nuthatch is common, and the presence of the somewhat elusive tree creeper is shown[4] by the numerous roosting holes scratched out of the bark of almost every specimen of *Sequoia gigantia*.

[1] By W. H. Thorpe, M.A., Ph.D.

[2] A more detailed account is the excellent study by D. Lack, *The Birds of Cambridgeshire* (1934). I am indebted to this, and to the reports of the Cambridge Bird Club.

[3] D. Lack, *op. cit.* p. 13.

[4] W. H. Thorpe, "The Roosting Habits of the Tree Creeper", *British Birds*, xviii, 20 (1924).

This roosting habit is of interest in that there is no native European tree that has a bark sufficiently soft to allow of it, yet, since the introduction of the *Sequoia* in 1853, the habit has become established throughout the British Isles. Where undergrowth is found, bullfinch, goldfinch and hawfinch breed, and the nightingale, lesser redpoll, and spotted fly-catcher also occur; while, most notable of all, is the recently established nesting of the black redstart in the centre of the town.

The trees of Cambridge also provide nesting sites for the rook, which is so common that the town may be described as one large rookery. While at Madingley Hall, a few miles to the west, is a rook roost which accommodates 15,000 or more birds in winter: here come the birds which feed within a radius of six to eight miles or more.

There are also starling roosts in the district, some accommodating as many as 120,000 birds, but these shift very considerably and the *Annual Reports* of the Cambridge Bird Club should be consulted for details.

The natural vegetation of the chalk upland with its occasional beech woods is found at Royston Heath, Newmarket Heath, and in very restricted portions of the Gogmagog Hills.[1] The characteristic birds here are the woodpigeon, skylark, meadow pipit, and corn bunting. Stone curlews nest regularly in small numbers, and quails breed in some years—mere remnants of their former vast hordes. In winter, bramblings frequent the beech woods in considerable numbers. Elsewhere, the chalk country is under crops, and birds are sparse, though large flocks of lapwing, golden plover, redwings, and fieldfares are a feature of the winter landscape.

East of Newmarket Heath, lies that great area of sands, gravels and boulder clay, the Breck country, occupying 400 square miles of Norfolk and Suffolk, and bordering Cambridgeshire.[2] Its barren sandy heaths and pine woods are characterised by stockdove, woodlark, nightjar, wheatear, stone curlew, and crossbill. There is also that curious inland breeding "race" of ringed plover which perhaps represents a relic of the littoral fauna of the old Fen Estuary. The Forestry Commission is however rapidly altering the aspect of much of this country,[3] and this close planting has had considerable effect upon the distribution of certain species.[4]

The last and the most characteristic type of country in the Cambridge district is, of course, the Fenland. By far the greater part of this area is now under cultivation, and the corn bunting, sedge warbler, reed bunting, and, more rarely, the corn crake, are among the characteristic species. The tree sparrow and magpie are also abundant—in unexplained contrast to their comparative scarceness south of the Cambridge-Newmarket road.

[1] See p. 57 above. [2] See p. 208 below. [3] See p. 217 below.
[4] D. Lack, "Habitat Selection in Birds", *Journ. Animal Ecology*, ii, 239 (1933).

In winter, besides great flocks of lapwing and golden plover, the black-headed and common gulls are numerous, while pink-footed geese are not infrequently found, particularly near Wisbech. Although there are patches of open uncultivated fen country at Fulbourn, Chippenham, Reach, Quy, and Burwell, almost the only remnant of undrained fen is at Wicken.[1] But, particularly because of the lack of reed beds and open water, the avifauna of Wicken Fen is only a fraction of what it once was. Gone beyond recall are pelican, crane, and spoonbill, that once inhabited the fens. Gone too, as breeding species, are Savi's warbler, bearded tit, black-tailed godwit, ruff, black tern, and the bittern. But the last three or four of these are still visitors to the district, and there is a possibility that some of them might be induced to return if conditions were made suitable. However, the Montagu's harrier and the short-eared owl still breed at Wicken in most years, and the grasshopper warbler is perhaps the most abundant and the most characteristic small bird of the Fen, while, at other seasons, marsh and hen harrier, peregrine falcon, merlin and common buzzard are occasionally seen. The existing open water attracts mallard, shoveller, teal, garganey, tufted duck, wigeon and the pochard, the first four as breeding species. But there is no doubt that the greatest need of Wicken as a bird reserve is the digging of a large mere and the encouragement of reed beds.

Finally, no summary of the ornithology of the district, however brief, would be complete without mention of the Cambridge Sewage Farm, two miles north of the town. Regular watching, mainly by members of the Cambridge Bird Club, has revealed an astonishing variety of passage birds, particularly of waders. Of special interest are the records of yellowshank, turnstone, curlew, sandpiper, Temmincks' stint, grey phalarope, and dotterel. Indeed, more wading birds have been recorded at the Cambridge Sewage Farm than at any other inland locality in Britain, and the observations carried on there have done much to discredit the theory that birds on inland migration follow definite routes such as the courses of rivers. All the observations in this district go to show that waders, when migrating, habitually fly at a considerable height and move on a broad front across country.

## REPTILIA

These include the common lizard (*Lacerta vivipara*), which seems to be local in distribution, but which is plentiful in Wicken Fen. The sand lizard (*L. agilis*) occurs about the Devil's Ditch near Newmarket, while the slow

[1] See pp. 45 and 50 above.

worm (*Anguis fragilis*) has been recorded by Prof. Stanley Gardiner from Wicken Fen. The grass snake (*Tropidonotus natrix*) occurs in suitable places, but it is very doubtful whether the viper (*Vipera berus*) can still be found in the County.

## AMPHIBIA

Apart from the common frog and common toad, which are prevalent throughout the County, the natterjack (*Bufo calamita*) occurs chiefly at Gamlingay, where its spawn is to be found in the shallow water of some of the clay pits. The edible frog (*Rana esculenta*), though once common, is now seldom found. The crested or warty newt (*Molge cristata*) is common in ponds and ditches, while the common newt (*M. vulgaris*) is very general in its occurrence. The palmated or webbed newt (*M. palmata*) seems to be confined to Quy Fen; at least, there are no records from other parts of the County.

## PISCES[1]

The sea lamprey (*Petromyzon marinus*) occurs in the River Nene and is sometimes caught above Earith. The river lamprey or lampern (*Lampetra fluviatilis*) is common in the Hundred Foot River, in the Ouse above Earith, in the Little Ouse, and in the Nene; and a number of lamperns was found in the Cam near Grantchester about the year 1927. The salmon (*Salmo salar*) is now only an occasional visitor. Trout (*S. trutta*) occur in the more rapid streams but are not very common. Pike, roach, dace, eels, minnow, rudd, tench, gudgeon, bleak, loach, perch, and miller's thumb, are all common. The grayling (*Thymallus thymallus*) is not indigenous but has been introduced into the River Lark. The chub (*Squalius cephalus*) is rather local and occurs near Cambridge in Byron's Pool. The silver bream (*Blicca bjoerkna*) and the bream (*Abramis brama*) occur commonly in the Fenland, while the Crucian carp (*Carassius carassius*) is apparently rare, and the common carp (*Caprinus carpio*), too, is not abundant. The spined loach (*Cobitis taenia*) occurs locally near Cambridge, and the burbot (*Lota lota*) is common in parts of the Fenland waters. The three-spined stickleback (*Gasterosteus aculeatus*) and the ten-spined stickleback (*Pungitius pungitius*) are both common: the latter occurs in fen ditches and lodes up to Lingay Fen above Cambridge. The flounder (*Platichthys flesus*) is frequently taken in the fenland rivers. Various marine fishes have been caught near Wisbech but, excepting the grey mullet, greater weever, and the dory, they have only been represented by single records

[1] From data supplied by H. E. Whiting, B.A.

## MOLLUSCA[1]

The neighbourhood of Cambridge is very favourable for Mollusca. The Fenland and the quiet waters of the Cam and its tributaries form a suitable habitat for many freshwater species; thus, among the Gastropods, in addition to five species of *Lymnæa*, no fewer than eleven members of the Planorbidae have been found within about a mile of Cambridge. *Theodoxus fluviatilis* (Lin.) and various other operculate forms also live in the Cam near the college bridges, while from Wicken as many as ten species of the Pelecypod genus *Pisidium* have been recorded. *Vertigo moulinsiana* (Dupuy), a scarce land snail restricted to marshy places, may also be found in Wicken Fen, and *Laciniaria biplicata* (Mont.) lives close to the river not far from Cambridge itself, although it is found in very few other places in the British Isles.

Upon the chalk hills to the south of Cambridge, on the other hand, xerophilous species are common; *Helicella virgata* (da Costa), *H. gigaxii* (Pfr.), and *Monacha cantiana* (Mont.) being especially abundant, the two former showing much variation; while about 5 miles south of Cambridge is found one of the very few British habitats of the large *Helix pomatia* (Lin.). *Helicigona lapicida* (Risso) occurs at Fen Ditton, but it is rare in Cambridgeshire, whereas its ally *Arianta arbustorum* (Lin.) is common and has even been known to find its way into the roof of King's College Chapel.

Slugs are not exceptionally abundant in the neighbourhood of Cambridge, but about eleven species have been found, and *Agriolimax reticulatus* (Müll.) and *Arion hortensis* (Fér.) are both very common, the former varying greatly in colour. Moreover, all the eleven British species of the Zonitidae, a family of snails related to some of the slugs, have been recorded from the district.

In gardens around Cambridge, *Trichia striolata* (Pfr.), *Helix aspersa* Müll., and other forms are abundant, and in the University Botanic Gardens six or seven exotic species have become established in the hot-houses. Excluding these foreign introductions, about 110 species of land and freshwater Mollusca are known to live in Cambridgeshire, as well as two or three brackish-water forms that occur in the north of the County, e.g. *Hydrobia ventrosa* (Mont.).

Most of these 110 species are also present in the local Pleistocene and Holocene gravels, but some of them have not been found in these deposits, including certain species that are now among the commonest in the neighbourhood, such as *Trichia striolata* (Pfr.) and *Monacha cantiana*

[1] By Hugh Watson, M.A.

(Mont.). On the other hand, the gravels contain several species that have not yet been reported alive in Cambridgeshire. Some of these, such as *Ena montana* (Drap.) and *Helicodonta obvoluta* (Müll.), still live in other parts of England; others, such as *Clausilia pumila* (Pfr.) and *Corbicula fluminalis* (Müll.), are now found alive only on the Continent; and one or two, such as *Helicella crayfordensis* Jackson, seem to be wholly extinct. In fact, the extensive river-gravel system of Cambridgeshire throws valuable light on the gradual modification of its molluscan fauna from middle Pleistocene times to the present day. It is possible, however, that further search will show more living species than are at present known; for although Cambridgeshire is rich in Mollusca, it numbers very few collectors who are interested in these animals.

## ARACHNIDA[1]

Records from the County comprise 245 spiders, 10 harvest spiders, and 7 pseudo-scorpions. The Acarina are not described. In collecting, the Fenland has deservedly received the greatest attention and Wicken Fen in particular. *Neon valentulus* Falc. (a small dark Salticid), *Maro sublestus* Falc. and *Centromerus incultus* Falc. (small black Linyphiids), are unknown elsewhere; *Zora armillata* Sim. has not been found elsewhere in Britain (this is a pale speckly Clubionid); *Maso gallica* Sim. has also been recorded only from Kent, *Entelecera omissa* Camb. doubtfully from Northumberland, and *Singa herii* Hahn. doubtfully from Berkshire.

May and June are the best months for collecting spiders at Wicken, and on a sunny day careful search will reveal several of the so-called rarities in abundance. Enclosed in silken cells in the fluffy heads of *Phragmites* will be found the handsome Salticid, *Marpessa pomatia* Walck. Running in the open, alongside the large velvety *Pirata piscatoria* Clerck and other Lycosids, will be seen the lighter coloured *Pardosa rubrofasciata* Ohl. Amongst clumps of hay-coloured grass will be found both the pale elongate *Tibelli* and another less common and also pale Thomisid, *Thanatus striatus* C.L.K. Most of the rarities must be sought for by grubbing at the roots of herbage or by turning over bundles of cut reeds: the black *Zelotes latetianus* L.K., the small rather pinkish *Clubiona neglecta* Camb., the large thick-set Lycosids *Trochosa spinipalpis* F. Camb. and *T. leopardus* Sund., the speckly Salticid *Sitticus caricis* Westr. and such small uncommon Theridiids and Linyphiids as *Crustulina sticta* Camb., *Theridion blackwallii* Camb., *T. instabile* Camb., *Taranucnus setosus* Camb., *Mengea warburtonii* Camb., *Gongylidiellum murcidum* Sim. and *Wideria melanocephala* Camb.

[1] By W. S. Bristowe, Sc.D.

Some of the College cellars provide a number of species. Rather damp cellars provide the largest fauna, but the long-legged and small bluish humpy-bodied Pholcid, *Physocyclus simoni* Berl., is exceptional in liking dry wine cellars such as those of King's and Trinity Hall. Since first adding this species to the British list in 1932, the present writer has discovered it in no less than nine counties, but, apparently, it has not been found elsewhere in Britain. Abroad, it is known in France only. The Pholcid, *Pholcus phalangioides* Fuess. is also present in Cambridge cellars, but its much larger size, different coloration and somewhat elongate body, easily distinguish it.

The only British Mygalomorph spider in the County, *Atypus affinis* Eich., has been recorded from Devil's Dyke, where its closed silken tube, like the finger of a glove, should be sought amongst vegetation on the sloping bank.

The recorded harvest spiders and pseudo-scorpions do not include any special rarities.

### INSECTA

Accounts of five of the major groups in this class are given below. These will serve to give an idea of some of the more noteworthy species that are to be found in the County. Good reference collections of all the major, and most of the smaller, orders of insects are contained in the University Museum of Zoology.

*HEMIPTERA (HETEROPTERA)*.[1] Cambridgeshire contains four main types of country, each of which possesses a distinctive Heteroptera fauna correlated with the associated flora and soil conditions. To the south lies the chalk; westwards are heavy clays; to the north is the drained alluvium of the fen basin; and in the east the boundary includes a small tract of the Breckland (see Fig. 56). The County list contains 256 species, out of 492 recorded for Britain. Drainage and agriculture have changed conditions, and in the Fenland at least the dominant species to-day are more typical of cultivated land than of marsh.

Although the fen basin was formerly an estuarine sea, no coast-loving species seem to have survived there, with the exception possibly of *Rhyparochromus praetextatus*, *Teretocoris antennatus* and *T. saundersi*, *Salda pallipes*, and a doubtful record of *Piesma quadrata*. On the other hand, the freshwater fauna may not have changed greatly; for although since the draining, standing water has diminished in extent, conditions in the habitat itself have probably remained fairly constant. Of the seventy-five species of water-bugs recorded for Britain, forty-seven have been found in Cambridgeshire. Most of them are widely distributed forms, charac-

[1] By Mrs M. D. Brindley.

teristic of the Northern Palaearctic region. An interesting species found at Wicken is *Glaenocorisa cavifrons*, which, apart from a larger darker form long known from the Scottish Highlands, has a restricted range in Britain. Four forms of *Notonecta* occur, while *Naucoris cimicoides* and *Ranatra linearis* are frequent in ponds. Cambridgeshire is not well provided with running water, but where it is found, as in the River Cam, *Velia currens* (Veliidae) and *Hygrotrechus najas* (Gerridae) are common. The five other Gerrids found in Cambridgeshire are inhabitants principally of standing water.

Turning to the land bugs, *Chartoscirta elegantula*, which in Britain is restricted to four counties only, is found at Wicken. Other species, recorded there and not elsewhere in the County, are *Hebrus ruficeps*, *Myrmedobia tenella*, *Pamera fracticollis*, *Teratocoris antennatus* and *T. saundersi*, *Adelphocoris ticinensis*, *Eurygaster maurus*, *Cyrtorrhinus geminus* and *C. pygmaeus*. *Chilacis typhae* occurs at Wicken and elsewhere where the reed-mace grows. *Doliconabis lineatus* is frequent on reeds, and *Oncotylus viridinervus* is found on *Centaurea* in fen pastures.

Characteristic species of the chalklands are *Calocoris roseomaculatus*, *Poeciloscytus unifasciatus*, and *Myrmus miriformis*. *Amblytylus affinis*, *Onychunemus decolor*, *Hoplomachus thunbergi*, and *Halticus apterus*, often occur in some numbers. *Eremocoris podagricus* has been recorded from near Royston, and two species of *Berytus* with *Metacanthus punctipes* are frequent, especially where the chalk bears hawthorn scrub. Where the fields are bounded by screens of *Pinus*, there appears an intrusive population of conifer-dwellers, such as *Gastrodes ferrugineus* and *Acompocoris pygmaeus*.

The claylands sometimes bear deciduous woodland.[1] As the Heteroptera here have been little studied, it is possible that additional species await discovery. Various Pentatomidae occur, such as *Eusarcoris melanocephalus*, *Palomena prasina* and *Gnathoconus albomarginatus*; while other characteristic species are *Macrotylus solitarius*, *Macrolophus nubilus*, and *Calocoris ochromelas*.

In the fruit-growing districts, the Capsids *Plesiocoris rugicollis* and *Lygus pabulinus* are pests of apple and currant. Other species of interest in the neighbourhood are *Reduvius personatus*, which is sometimes taken at dawn in the town; and the bird and bat parasites, *Cimex columbarius*, *Oeciacus hirundinis* and *Cimex pipistrelli*, which were first described by Jenyns from Cambridgeshire.[2]

*LEPIDOPTERA*.[3] For at least two centuries Cambridge and its neighbourhood have been celebrated for their butterflies and moths, and although some of the more interesting species have now become extinct,

[1] See p. 52 above.
[2] See p. 60 above.
[3] By J. C. F. Fryer, O.B.E., M.A.

the fauna is still a remarkable one. Perhaps the best way to give the reader a brief introduction to it is to consider rather the different types of habitat exhibited by the country near Cambridge than to attempt any description of the different species.

First, Cambridge itself deserves mention, since its old walls are the haunt of a special race of a moth (*Bryophila muralis*), of which the typical form is largely confined to the southern and western coasts. The species has never established itself in the adjacent villages (even when introduced) but, in Cambridge town, it seems able to survive the changes of modern times. The moth appears in August and the larvae feed on the algal growth on old walls, like those of its common relative, *B. perla*.

Leaving Cambridge, the most important habitat is that of the Fenland. Most of this area is intensively cultivated, but fen species persist in the dykes and clay pits; instances are the local "Wainscot" moths, *Leucania obsoleta*, *Senta maritima* and *Nonagria arundinata*, found where the common reed is left uncut, and the marsh carpet (*Cidaria sagittata*), a rare and local species occurring in dykes (also fens) where its food plant—meadow rue—grows. A few areas remain in a more primitive condition, the fens of Wicken and Chippenham in Cambridgeshire being the most famous. Wicken, best known for the swallow-tail butterfly, is also the haunt of many interesting species, such as the reed leopard (*Macrogaster castaneae*) and the marsh moth (*Hydrilla palustris*), the latter a very rare insect with the habit of flying chiefly between midnight and dawn. The Dutch large copper butterfly, very closely resembling the extinct English large copper[1] is being re-established at Wicken.[2] Chippenham somewhat resembles Wicken in its Lepidoptera, but the swallow-tail is not found there, although some other species occur more abundantly—as, for instance, the Noctuid *Bankia argentula*, elsewhere in the British Isles almost confined to Killarney.

The higher land immediately bordering the Fenland, and also the higher parts of the Isle of Ely, support three characteristic Tortricid moths (*Phtheochroa schreibersiana*, *Pammene trauniana*, *Laspeyresia leguminana*) and, in addition, one of the scarcest of British "dagger" moths (*Acronycta strigosa*), which elsewhere has only been found near Tewkesbury, a remarkable distribution in view of the universal occurrence of hawthorn, its food plant.

Next to the Fenland, the most important area is that of the Chalk in the

[1] See p. 188 below.

[2] Woodwalton Fen, not far from the site of the former Whittlesea Mere, in Huntingdonshire, contains some of the fen species found at Wicken, and also another local Noctuid *Tapinostola extrema*, at one time thought to be extinct. This fen is best known for the successful re-establishment of the Dutch large copper butterfly a race very closely resembling the extinct English large copper.

south of the County. Here, most of the typical chalk Lepidoptera are found, e.g. the chalk hill blue, which occurs in an interesting race at Royston. The best localities for seeing the chalk Lepidoptera are the Devil's Dyke, the Fleam Dyke and the Roman Road.

Two other areas, each with a different fauna, are also accessible from Cambridge. Extending to the County borders in the south-east is the Breckland of Norfolk and Suffolk, which has a very characteristic fauna. Such species as the Noctuids *Dianthoecia irregularis*, *Agrophila trabealis* and the Geometrid *Lithostege griseata* are found nowhere else in the British Isles, while some species otherwise largely confined to the seashore occur there also.

At the opposite side of the County, just across the border into Huntingdonshire, another distinct fauna is found, that characteristic of oak woods growing on clay, the most characteristic species being the black hairstreak, which in Britain has a very restricted distribution. Monks and Warboys Woods are typical of this type of country and are those best known to lepidopterists.

*COLEOPTERA.*[1] The following account of beetles deals only with the rarer species. Many species thought, in 1904, to occur only in Cambridgeshire have since been taken elsewhere.

*Aleochara fumata* Gr. (Fowler—*A. brevipennis* Gr. var. *curta* Sahlb.): taken by the late G. C. Champion at Soham; since been taken by the writer in the New Forest and Windsor Forest. *Rhantus adspersus* J.: taken in profusion by Charles Darwin, but is now apparently extinct in Cambridgeshire. *Trichopteryx championis* Matt., *Ptilium caesum* Er. and *P. incognitum* Matt.: all from Wicken Fen, where they have not been taken again, nor have they been recorded from any other locality. *Cryptophagus schmidti* Strm.: two specimens taken by G. C. Champion in Wicken Fen, and one by the late E. W. Janson at Whittlesea, remained unique, until it was rediscovered by the late Miss F. J. Kirk and the writer in Burwell Fen. *Cryptocephalus primarius* Har.: a single specimen was taken by the late Dr Power on the Gogmagog Hills; not since taken, until discovered by R. O. Richards, and also taken by J. Collins, near Oxford. *Tychius polylineatus* Ger.: introduced by Crotch on a specimen taken by himself at Cambridge, about 1863; was retaken in fair numbers by the late Hereward Dollman on the downs at Ditchling, Sussex.

The following species have not been found in Cambridgeshire for many years: *Pterostichus aterrimus* Pk.: formerly common in the Fens; some years ago now Sir T. Hudson Beare took a single specimen in Norfolk, but Mr Bullock has taken it in numbers at Killarney in recent ears.

[1] By H. St J. K. Donisthorpe, F.R.E.S.

*Graphoderes cinereus* L.: has not been taken in the Cambridgeshire fens for very many years; F. Balfour-Browne took a fair number in Norfolk, over twenty years ago.

Of the species Fowler considers to have disappeared before the draining, one may mention: *Trechus rivularis* Gyll.: thought to be extinct. The late A. J. Chitty and the writer took it sparingly in cut sedge bundles at Wicken Fen in 1900. Since then, however, it has been taken again in some numbers by several collectors. *Dytiscus dimidiatus* Berg.: considered to have become exceedingly rare; in 1899 and 1900 the writer took it not uncommonly in Wicken Fen. *Oberea oculata* L.: also considered to have disappeared; was plentiful in 1898 and 1900, and was also taken on other occasions by Beare, Bouskell and the writer. I believe, however, it has got more scarce again. *Lixus paraplecticus* L., also supposed to have disappeared, was found by Mr F. Bouskell and the writer in fair numbers, in 1894, and appeared to be spreading. This also I believe is getting scarce again.

Space allows only brief notes on a few species from the different sections of Coleoptera.

In the Geodephaga or ground beetles, *Ophonus obscurus* F., which occurs at the foot of the Devil's Dyke near Swaffham, appears to be almost confined to Cambridgeshire, though the writer has taken it at Abbotsbury in Dorsetshire. *Chlaenius holosericeus* (*tristis* Schal.) was formerly recorded at Fen Ditton in 1827, and by Charles Darwin near Cambridge. Dr Power took it in Burwell Fen, but it has not since been taken in Cambridgeshire. The beautiful *Panageus crux-major* L. occurs sparingly under horse-cut sedge; it used to be more plentiful formerly.

The water beetles are well represented and we have already dealt with the most interesting species. The most noteworthy of the Hydrophilidae is the large *Hydrophilus piceus* L., which used to be common under water lilies in the Wicken Poor's Fen; it is scarcer now. The very rare *Spercheus emarginatus* was taken by Prof. Babington in Burwell Fen; it has not occurred in Cambridgeshire since.

It is hard to choose which species to mention out of the very large number of Staphylinidae recorded. *Microglossa marginalis* Gr., once recorded as British by Crotch (a single example near Cambridge), has since been taken in birds' nests in many other counties. The very rare *Schistoglossa viduata* Er. has recently been retaken in Cambridgeshire by E. C. Bedwell in Wicken Fen. Passing on to the Clavicornia, *Silpha tristis* Ill. is found on paths and at roots of grass at Wicken Fen, and the much rarer *S. nigrita* Creutz. was taken by Dr Power on the Gogmagog Hills.

*Copris lunaris* L., in the Lamellicornia, was recorded by the Rev. L. Jenyns as plentiful in 1828 in a field near Melbourn; but it has not been taken in Cambridgeshire since. Of the Serricornia, the last specimen recorded for the very rare *Ludius ferrugineus* L. was taken on a poplar by the Cam about 100 years ago. It has since been taken and bred in Windsor Forest by the late Miss Kirk and the writer in some numbers in recent years. *Platycis minutus* J. was discovered by the late G. H. Verrall in Chippenham Fen in 1898, and has been taken there by the writer and other collectors.

Of the longicorns, *Oberea oculata* L. has already been dealt with. The large *Saperda carcharias* L. used to be abundant both in Wicken village and the Fen some thirty years ago. I understand it is much scarcer now. *Agapanthia lineaticollis* Donov. may still be swept off thistles in the fens.

In the Chrysomelidae, *Chrysomela graminis* L., which used to be abundant on water mint in various parts of the fens, is much scarcer now I am told, and restricted to small local patches. A number of the Donaciae are found, and the rare *D. dentata* Hoff. used to be common on various water plants. *Adimonia oelandica* Boh., a very rare species, was found by the late Mr Blatch in number in Wicken Fen in 1878; it has not occurred since.

Among the Heteromera, the most interesting species are *Cteniopus sulphureus* L., a coast species, not uncommon by sweeping in Wicken Fen. *Lytta vesicatoria* L., the "blister beetle", was originally recorded from the Gogmagog Hills, where it was found again by the writer in 1901. It is sometimes abundant at Newmarket, etc., on privet hedges. *Anthicus bifasciatus* Rossi (new to Britain) was discovered by the Fryers in manure heaps round Chatteris; subsequently taken by Williams in a manure heap at Wicken; and by the late Miss Kirk and the writer in Burwell Fen; all in numbers. It has since been found in Oxfordshire (J. J. Walker), etc. The Rhynchophora, or weevils, are abundant. *Ceuthorrhynchus angulosus* Boh., a local and rare species, was taken by J. C. F. and J. H. Fryer on *Stachys* and *Galeopsis* at Chatteris and Somersham (they also took the rare Halticid *Dibolia cynoglossi* Koch. on *Galeopsis* at Somersham). The curious *Lixus paraplecticus* L., which occurs on *Sium latifolium* in Wicken Fen, has already been mentioned. It is covered with a yellow dust, which is renewed in life. The local *Dorytomus salacinus* Gyll. can be beaten off sallow bushes in Wicken Fen, etc.

*DIPTERA.*[1] Our knowledge of the distribution of species of Diptera is still very incomplete. Intensive collecting, even in a county such as Cambridgeshire, could not fail to produce a number of species previously unrecorded (or even undescribed). We know still less of the changes that

[1] By J. E. Collin, F.R.E.S.

may have taken place in the fauna but it is interesting to note that Jenyns[1] recorded the occurrence of three of the larger species of Diptera, the Tabanids, *Tabanus bovinus* L. at Ely and Bottisham and *Atylotus rusticus* F. at Cambridge, and the large Tachinid *Peleteria nigricornis* Mg. on the Devil's Dyke. The identity of the last two can be proved by an examination of Jenyns' specimens, but none of these three has been since taken in the County. An interesting case of the reverse condition is that of the handsome Trypetid *Anomaea permunda* Harr. (*antica* Wlk.). Formerly regarded as a rarity it is now abundant, at least locally, and is freely bred from hawthorn berries gathered near Cambridge. Recently it was found in such numbers on the windows of a house on the outskirts of that town as to be considered a "pest".

The physical features of a county are always of primary importance in connection with insect life. The Fenland in the north of Cambridgeshire harbours many species not found in the drier strip of the Chalk to the south. This, and the clay strip of south-eastern Cambridge, the river valleys, and the fringe of Breckland on the east, all provide characteristic species (see Figs. 29 and 56).

Of particular interest is Wicken Fen, where intensive collecting might well produce species unknown elsewhere in the whole country. It was here that the author found the Chloropid *Lipara similis* Schin. which attacks the growing point of the reed (*Phragmites communis*) without doing much apparent damage, while its close relative *Lipara lucens* Mg. causes a large gall-like swelling. Here also was found in 1935 and 1936 the tiny midge *Pterobosca paludis* Mcfie. sucking the juices from the wing-veins of dragonflies. An unexpected capture in 1936 was that of the rare Syrphid *Myiolepta luteola* Gmel.; but some hollow tree on the outskirts of the fen must have been harbouring this species for many years, in the same way that the pollard willow trees at Upware probably accounted for the capture years previously of *Xylomyia marginata* Mg. The occurrence of the rare *Odontomyia angulata* Pnz.—a marshland species—is not surprising.

Chippenham Fen, not very far from Wicken, is almost surrounded by woods and plantations and therefore possesses a somewhat different fauna. This is the home of the uncommon Syrphids *Chilosia nebulosa* Verr., and *Sphegina kimakowiczi* Strbl., and here a specimen of the giant Pipunculid *Nephrocerus flavicornis* Ztt., and the interesting Conopid wasp parasite *Brachyglossum* (or *Leopoldius*) *signatum* W. have been taken. Here are also to be found a few rare Trypetids such as *Spilographa abrotani* Mg., *Rhacochlaena toxoneura* Lw., and *Oxyphora corniculata* Fln., and other Acalyptrates such as *Ochthiphila coronata* Lw., *spectabilis* Lw., and *elegans* Pnz., and

[1] See p. 60 above.

*Chymomyza costata* Ztt. Chippenham Fen is also at present the only locality (in addition to Spain) where the very tiny Dolichopodid described by Strobl as *Micromorphus albosetosus* is known to occur.

On the chalkland of the south, the Devil's Dyke and Fleam Dyke are excellent localities for the downland species associated with chalk. Here, rare Tachinids may be obtained such as *Lydella angelicae* Mg., *Demoticus plebeius* Fln., *Zophomyia temula* Scop., *Neaera albicollis* Mg., and *Ocyptera interrupta* Mg., as well as many species of *Sarcophaga*. It was here that the new Trypetid *Trypeta* (*Ceriocera*) *microcera*, recently described by Dr Hering of Berlin, was found by Mr G. C. Varley living in the stems of *Centaurea scabiosa*, while many other interesting Acalyptrates occur.

The clayland woods (see Fig. 29) begin at Woodditton, and their associated insects are naturally different from those of the rest of the County. Among the Syrphids, *Chilosia maculata* Fln. is common on the wild onion; *Platychirus tarsalis* Schum. is common on the flowers of *Geum rivale*, while the rare *Chilosia pubera* Ztt. and *fasciata* Egg. have also been taken. The peculiar plant *Paris quadrifolia* is the host plant of the Cordylurid *Parallelomma paridis* Her. Interesting Tachinids such as *Camplyochaeta praecox* Mg., *Actia nigrohalterata* Vill., and *Blepharomyia amplicornis* Ztt. have occurred; while the rare *Xysta cana* Mg. has been taken at Kirtling not far away. Among the Acalyptrates, the rare *Acartophthalmus bicolor* Old. was once found in Woodditton Wood sitting on dead twigs at the bottom of a dense thicket.

An account of the Diptera of Cambridgeshire would not be complete without mention of some of the captures of the late Francis Jenkinson of Cambridge. Particularly interesting was the occurrence of two very little known Tachinids (*Stenoparia monstrosicornis* Schin., and *Helocera delecta* Mg.) in his garden, but he also found at Cambridge the large Pipunculid already mentioned—*Nephrocerus flavicornis* Ztt.—and the Drosophilid *Acletoxenus formosus* Lw., together with many other good species too numerous to mention including (in 1901) the rare Tachinid *Stomatorrhina lunata* F., which is probably only an occasional visitor to this country. Finally, Jenkinson and other Cambridge entomologists have proved that many of the rare species associated with rotting wood are to be found in and about the very old trees of the College Gardens and along the Backs. These include such species as the Syrphids *Mallota cimbiciformis* Fln., and *Pocota apiformis* Schrnk., the four species of the Dolichopid genus *Systenus*, as well as many Anthomyids and Acalyptrates which frequent sappy exudations.

*HYMENOPTERA.*[1] Although our knowledge of the Hymenoptera of Cambridgeshire is in advance of that of most counties, it is considerably

[1] By G. J. Kerrich, M.A.

behind our knowledge of, say, the Coleoptera; and an attempt to assess the hymenopterous fauna of the County must, in the main, be regarded as preliminary.

The sawfly fauna[1] should be very rich, considering the abundance and variety of willows, grasses, and horse-tails, the most characteristic food plants of these insects; and sawflies certainly are numerous in individual species, particularly in the fen country. Mr Benson collected them energetically during his student days, but they have since received very little attention; and, probably, the known total of 172 species could be nearly doubled. Three Pamphiliidae, five Cephidae, and five Cimbicidae are known: a Cimbicid larva sometimes rewards a search on willows. *Xiphidria prolongata* Geoffr. has several times been taken; and three Siricidae are known, as are both their parasites, *Rhyssa persuasoria* L. (Ichneumonidae) and *Ibalia leucospoides* Hochenw. (Cynipoidea). Of the Tenthredinidae, the occurrence of the rare *Ametastegia albipes* Thoms. is especially interesting.

For Aculeata, the County is much less favourable. On account of the absence of sandy ground, except along two parts of the County boundary, the numerous sand-living species are absentees or strays. The bees are the group least affected; Adrenidae, in particular, are well represented, and many of them visit willow flowers in the spring. Bumble-bees are conspicuous both in and around the fens and in College gardens. An interesting fen bee is the little *Hylaeus pectoralis* Först., which nests in old reed galls of the fly *Lipara lucens* Mg. Its parasite is *Gasteruption rugulosum* Ab. The best distributed ant is *Acanthomyops flavus* Fabr. The hornet, *Vespa crabro* L., nests in old willows and old cottages. *Cleptes semiauratus* L., a sawfly parasite, is fairly common; and the Chrysididae are well represented. Several of these latter belong to the interesting fauna which frequents old posts, including species nesting therein, and their parasites. There are many such posts along the approaches to Wicken Fen, and these have frequently been studied. The commonest species is *Trypoxylon figulus* L.; *Sapyga clavicornis* L., and the rare *Cuphopterus confusus* Schulz, recently discovered there by the late H. P. Jones, may also be mentioned. The species total of Aculeata for the County, excluding the Gamlingay district, is 236.

Of the gall-wasps, only those living on *Centaurea* and *Rosa* have been seriously studied. *Isocolus scabiosae* Giraud and *fitchi* Kieff., *Rhodites spinosissima* Giraud and *mayri* Schlechtd. have not been found; but the other known British species occur in the County, as do the two *Periclistus*

[1] I wish to thank Mr R. B. Benson for giving me access to the manuscript of his section on sawflies for the account in the *Victoria County History of Cambridgeshire*.

spp., inquilines in *Rhodites* galls. *Diastrophus rubi* Bouché and *Xestophanes potentillae* Vill. are known; but the oak fauna is almost untouched. The commonest Figitidae are known, but not much can be said of the other parasitic forms.

Of Ichneumonidae, 452 species are known, a number of which were first British records. Many species are attracted to the flowers of Umbelliferae, which are conspicuous along roadsides and in meadows. *Aritranis carnifex* Grav., *Epiurus melanopygus* Grav., and *Diblastomorpha bicornis* Boie., are characteristic fen species; also *Hemiteles balteatus* Thoms., known in Britain only from Wicken Fen and the Norfolk Broads. Tufts of *Deschampsia caespitosa*, in fens and poor wasteland, provide winter quarters for the females of numerous species. 180 Braconidae have been identified, the majority by the late G. T. Lyle. Only 28 Serphoidea are known; and the list of 60 Chalcidoidea is almost entirely composed of species bred in the course of general biological work.

*OTHER ORDERS*. Among the smaller orders a beginning has been made by Dr C. H. N. Jackson with the *Collembola*. From among a total of over 150 British species of this order, some 55 have been found in the County, where they have been mostly collected in Wicken Fen and near Cambridge. Among the *Orthoptera*, several of the rarer species have not been found for many years. The most notable recent record is that of the great green grasshopper (*Tettigonia viridissima*), which has been found at Madingley by G. C. Varley. The *Ephemeroptera* have been but little collected, and records of only 13 species are apparently known. Out of a total of 44 British species of *Odonata*, 27 kinds have been found in Cambridgeshire; most of the recent records are due to the late W. J. Lucas and J. Cowley. About half the recorded British species of *Psocoptera* (booklice) have been found in Cambridgeshire by R. M. Gambles and others. Although a few rare species occur in Wicken Fen, there are none that are peculiar to the County. Knowledge of the *Hemiptera-Homoptera* is still scanty, and much work needs to be done on this suborder before an adequate idea of its representatives can be ascertained. In the *Neuroptera*, some 34 species out of a total of 57 British forms have been recorded. The snake flies *Raphidia xanthostigma* and *R. maculicollis* occur in woods, while *Sisyra fuscata* may be found around Cambridge and at Wicken Fen, and probably elsewhere, along with the freshwater sponges with which it is associated. The most notable member of the order is the very rare *Psectra diptera*, an example of which was taken in Wicken Fen in 1934 by H. Donisthorpe. Of the *Mecoptera*, all three British species of *Panorpa* occur, *P. cognata* being recorded from Fleam Dyke. The *Trichoptera* or caddis flies are well represented, especially in Wicken Fen. The most interesting records are perhaps those of the very local *Limnophilus decipiens* and of *Agraylea pallidula*, which were found in the Fen by M. E. Mosely in 1926: the last-named had only been taken once previously in Britain. Some 63 species representing 11 families have so far been found, but there is much scope for futher work on this order.

## MYRIAPODA

According to E. B. Worthington, 28 species have been found in Cambridgeshire, and those requiring special mention include *Glomeris marginata* and *Polyxenus lagurus*. The last-named has been found in damp timbers around Cambridgeshire and Wicken. The difficulties attending their identification and the paucity of reference collections probably account for the scanty attention given to the "myriapods" in Britain.

## CRUSTACEA[1]

Considering that Cambridgeshire is so essentially an inland county, its crustacean fauna is truly remarkable. This is represented by no less than 76 genera and 166 species. Quite recently, the fauna of Wicken Fen has been investigated fairly thoroughly, but previously to that a great deal of work had been done by Brady and Robertson as early as 1870. Wicken of course represents a part of the original Fenland and possesses its own interesting fauna. There is also the remains of an interesting salt marsh with brackish water fauna to be found at Wisbech.

Among the higher Crustacea are records of *Carcinus maenas* (Pennant) and *Palaeomonetes varians* (Leach). *Chirocephalus diaphanus* Baird is also recorded from Bottisham Park, while there are several records of *Niphargus* or the blind well-shrimp, though it is quite certain that several of Gilbert White's records of the spring keeper refer to *Niphargus* and not to *Gammarus*. The higher Crustacea, with the addition of *Chirocephalus* and *Argulus*, are represented by 18 genera and 24 species.

The Cladocera are represented by 21 genera and 40 species. Most of them come from Wicken Fen, and a few of these are rare or of exceptional interest. *Macrothrix hirsuticornis* Brady and Norman, recorded by W. A. Cunnington, is a rare species, and so are *Acroperus angustatus* Sars and *Alona tenuicaudis* Sars, both recorded from Wicken Fen by P. M. Jenkin. *Anchistropus emarginatus* Sars, recorded by A. G. Lowndes from Wicken Fen, was at one time considered the rarest species of Cladocera in the British Isles, but it is really fairly common. *Polyphemus pediculus* L. is very abundant on Wicken Fen, but it does not occur in many of the midland counties.

[1] By A. G. Lowndes, M.A.

The Copepods are represented by 5 genera and 37 species, but it is safe to say that with the possible exception of Wicken Fen the group has hardly been touched. It is certain that a careful investigation of Whittlesea would greatly add to the number of species. Of the recorded species, that of *Cyclops gigas* Claus is the most important, since it is the only authentic record in the British Isles.

The Ostracods are or were till quite recently a sadly neglected group of Crustacea and yet they should be of considerable interest. In Cambridgeshire, the ostracod fauna is remarkable, no less than 32 genera and 65 species being recorded. *Siphlocandona similis* Baird is a rare species, but occurs fairly abundantly on Wicken Fen. The genus is not recorded outside the British Isles. *Prionocypris olivacea* (Brady and Norman) is a rare species, but it was found in large quantities quite recently at Ashwell (in Hertfordshire just outside the County) by P. F. Holmes. The genus *Pseudocandona* is also recorded from Wicken. This is an important record, for there has probably been more confusion over the species *P. pubescens* (Koch) than over any other species of freshwater ostracod.

The sperms of ostracods are the largest among the whole animal kingdom, and moreover they are highly mobile. In many species, and even in many genera, males are unknown and yet in those cases where males do occur the male genital organs are of a highly complicated type. There is on record one species, *Herpetocypris reptans* Baird, which was known to breed entirely by parthenogenesis for eleven years. The males of this genus are unknown while the females still retain spermatheca with ducts of relatively enormous length, presumably for the reception of these giant sperms, and yet it is pretty certain that the sperms ceased to exist long before the Tertiary period. Holmes has recorded a second species of *Pseudocandona* from Lake Windermere which promises to throw a considerable amount of light on this obscure subject.

There is yet another record which should be of great interest. *Hemicythere villosa* Sars is recorded from the River Cam, where it is quite abundant on occasions. It can be traced right out to sea and is recorded by Sars from some of the deepest fiords of Norway. The same species is also recorded in the fossilised state from the Pre-tertiary period.

## HIRUDINEA

W. Ambrose Harding records nine out of the eleven species of British freshwater leeches in the County. The medicinal leech (*Hirudo medicinalis*) has apparently disappeared from Cambridgeshire many years ago. Among the more local species, *Theromyzon tesselata* and *Hemiclepsis marginata* sometimes occur in numbers: the first-mentioned is stated to live upon water-fowl and the latter is at least partially a fish parasite. Very little definite information exists, however, with regard to their hosts.

### TURBELLARIA

Five species of flatworms occur in ponds and ditches, and two species are to be found in running water. These latter, viz. *Planaria alpina* and *Polycelis cornuta*, occur in springs where the temperature is low and varies little throughout the year. They are regarded as relics of a former glacial fauna which once populated the County.

### PORIFERA

The two freshwater sponges, *Ephydatia fluviatilis* and *Spongilla lacustris*, are prevalent. According to G. P. Bidder, the common species, *E. fluviatilis*, is to be found on wooden piles, lock gates, or under floating wood, in the waters around Wicken. *Spongilla* prefers deeper and moving waters, and may be found growing up from the bottom of the River Cam along the Backs at Cambridge.

CHAPTER SIX

# THE ARCHAEOLOGY OF CAMBRIDGESHIRE

Edited by J. G. D. Clark, M.A., PH.D.

(With contributions by J. G. D. Clark, T. C. Lethbridge, and C. W. Phillips)

CAMBRIDGESHIRE IS FAMOUS AMONG ARCHAEOLOGISTS FOR the distribution studies of Sir Cyril Fox. His book, *The Archaeology of the Cambridge Region*, has had a widespread influence, and it covers the extensive material housed in the University Museum of Archaeology and Ethnology, up to the time of its publication in 1923.

Since 1923, the most important work on the upland has been the dyke and cemetery excavations carried out by the Cambridge Antiquarian Society, which have served to place Cambridgeshire in the forefront of Anglo-Saxon studies. In the Fenland, the excavations sponsored by the Fenland Research Committee, with the assistance of the Percy Sladen Memorial Fund, have thrown a flood of light on the relation of successive phases of human settlement to the geographical evolution of the Fenland basin. A fuller survey of the prehistoric archaeology of the County, complete with references, will be found in the forthcoming *Victoria County History of Cambridgeshire*.[1]

The close connection between human settlement and land movement is brought out by Figs. 18–21. The most striking feature of these is the density of settlement in the southern fens during the Bronze Age, and the sparseness of settlement during the Early Iron Age. This change is certainly to be connected with the post-glacial subsidence of the area. In Romano-British times, the distribution of settlement was similar in broad outline to what it had been during the Early Iron Age, with the important exception that the silt fens in the north of the County and in south-eastern Lincolnshire were then intensively cultivated (see Fig. 47). This may have been due to a minor phase of re-elevation, but it may have been facilitated by the superior technical ability of the Romans. Finally, in Anglo-Saxon times the silt fens ceased to be cultivated. This change was due partly, perhaps, to the breakdown of drainage or defensive works, but also to a further slight subsidence. The emptiness of the peat fens at the close of this period is emphasised by the map of Domesday villages (Fig. 22).

[1] I am indebted to the Editor (Mr L. F. Salzman) for permission to use this material in the preparation of this chapter.

## THE PALAEOLITHIC AGE

The earliest certain traces of man in Cambridgeshire consist of flint implements incorporated in deposits dating from the late Prof. J. E. Marr's "Period of Aggradation", when gravel accumulated on the flood-plains of rivers in the southern part of the County while the northern portions were submerged beneath the sea.[1] The implements include hand axes of Acheulian type as well as Clactonian and Levalloisian flake tools. The only site in the County at which Lower Palaeolithic implements have been obtained from a well-studied geological section is the famous Travellers' Rest Pit[2] at Cambridge itself; unfortunately the pit has recently gone out of use and the section is no longer visible. Within the borough of Cambridge, also, the Lower Barnwell Village beds have produced implements at Chesterton and Barnwell. Large numbers of unabraded flake and core implements have been obtained at different times from the gravel ridge at Upper Hare Park, Swaffham Bulbeck. A few stray implements have come from the Granta Valley near Hildersham and Linton, while others from Girton, Oakington, and Willingham, mark the course of an extinct river that once flowed from the neighbourhood of Trumpington towards Earith. The gravel spread in the Kennet-Kentford area is known to have yielded many palaeoliths (many of them from over the Suffolk border), but precise information is lacking. The same applies to many finds from the fen islands (e.g. from "Shippea Hill" near Ely). A few flake implements of somewhat doubtful affinities have been obtained from the March gravels, which at this time were situated on the coastline.

No Upper Palaeolithic sites have so far been located in the County, but a few stray flints may point to their existence in the neighbourhood—notably an angle burin from Wicken which exhibits a remarkable dark green-and-white mottled patina. According to Marr, these Upper Palaeolithic flints should be contemporary with a "Period of Erosion", when the sea coast lay far out beyond the Dogger Bank, and when the rivers of Cambridgeshire were eroding their banks.

## THE MESOLITHIC AGE

During Mesolithic times Cambridgeshire, in common with southern Britain as a whole, began to undergo a progressive, though not uninter-

[1] Although closely bordering areas of intensive Palaeolithic research, Cambridgeshire does not occupy a prominent place in this field. The best documented finds can be seen in the Sedgwick Museum of Geology, where the admirable index catalogue compiled by the late Prof. J. E. Marr can be consulted. See also the papers by J. E. Marr in *Quart. Journ. Geol. Soc.* lxxv, 210 (1920), and *ibid.* lxxxii, 101 (1926).

[2] See p. 16 above.

rupted, subsidence, which was not fully accomplished until the Early Iron Age. The initial stage in this process was marked by the submergence of the North Sea "moorlog", from which the prong of a Maglemosian fish spear has been obtained, some 25 miles from the Norfolk coast. Part of a very similar specimen was found many years ago in the Royston district, probably from a low-lying site in the Cam Valley. Microlithic industries of Tardenoisian aspect have been found at Fen Ditton, at Chippenham, and on several sandy hillocks in the Ely fens. Excavations on the flanks of one of these hillocks at Peacock's Farm, near Shippea Hill station, revealed evolved Tardenoisian flints stratified in the lowermost peat bed underlying the fen clay and at a depth of some 17 ft. *below* mean sea-level (Newlyn). These flints were of an industry previously well known from the sand dunes between Wangford and Lakenheath, Suffolk. It is thus established that Cambridgeshire, by the close of Mesolithic times, was still at least 30 ft. higher in relation to the sea than it is to-day. Pollen analysis shows that the Late Tardenoisian industry immediately antedates the change-over from pine to alder dominated woods, which marks the Boreal-Atlantic transition in this area.[1]

## THE NEOLITHIC AGE

If stray finds of flint implements be excepted (and it is no longer possible in this country to assign any single type exclusively to this phase), there is very little material evidence for a Neolithic settlement of Cambridgeshire. The pottery obtained from the Peacock's Farm excavations, overlying the Late Tardenoisian level, shows, however, that the area was affected by the Neolithic "A" (Windmill Hill) culture, while the level at which it was found (*minus* 15 ft. O.D.) indicates that the subsidence was still at this period far from complete. It is likely that the Neolithic "A" culture spread to the Essex coast and the fen basin by direct overseas movements; but the long barrow at Therfield Heath, Royston, on the line of the Icknield Way, suggests that influences did move up the chalk belt from Wessex, although the absence (with one possible exception in Norfolk) of long barrows from the rest of East Anglia seems to indicate that such influences were unimportant.

No pots decorated in the "A2" style have yet been found in the County, but the recent discovery of a complete bowl in Mildenhall Fen, only a short distance over the Suffolk border, suggests that such finds are not unlikely in the future.

Nor, despite its proximity to the type site (Peterborough), can Cambridgeshire yet show any certain traces of the Neolithic "B" culture.

[1] See p. 18 above.

## THE BRONZE AGE

Although but little is yet known of the economy and dwelling sites of the Bronze Age inhabitants of Cambridgeshire, sufficient stray finds (mainly metal objects, but also a few pots) have been made to give some idea of the areas settled at this time (see Fig. 18). The densest zone of settlement was the fen margin between Cambridge and Isleham. Fen islands, such as March, Manea, Chatteris, Littleport, Ely, and Stuntney, have also yielded many finds; so have flat fens like Burnt Fen, Wilburton Fen, Grunty Fen, and the Chatteris-Mepal fens. On the upland, in the south of the County, Bronze Age finds are more or less limited to the chalk belt, the neighbouring areas of Boulder Clay, Gault, and Kimeridge Clay being virtually empty.

One explanation for the relative density of settlement in the Fenland during the Bronze Age is to be found in the natural conditions prevailing at this time. It is known, from excavations at Shippea Hill, that during the Early Bronze Age the fen basin was still several feet (at least 15 ft.) higher in relation to the sea than it is to-day. This was a height quite sufficient to affect profoundly the possibilities for settlement in such a low-lying area by prehistoric man. Conversely, as subsidence set in again, settlement tended more and more to move out of the fen basin. Already by the Late Bronze Age, half the hoards, and more than a third of the loose finds, come from the higher land in the southern part of the County: and by the Early Iron Age the evacuation was almost complete. The gradualness in the drift of population seems to discount the explanation that it was due to sudden economic change, and, in particular, to the introduction of a more intensive type of agriculture, which in this area can hardly have occurred before the Early Iron Age.

According to Sir Cyril Fox, the following stages can be recognised in the local Bronze Age:

| | |
|---|---|
| Transitional | 2000– 1700 B.C. |
| Early Bronze Age | 1700– 1400 B.C. |
| Middle Bronze Age | 1400– 1000 B.C. |
| Late Bronze Age | 1000–5–400 B.C. |

*Transitional and Early Bronze Age.* The earliest metal forms are rare in Cambridgeshire, comprising only four flat axes without expansion of the cutting-edge, and one round-heeled flat riveted dagger. Objects of the full Early Bronze Age, on the other hand, are plentiful, and include 17 flat axes with expanded cutting-edges, 23 flanged axes, two spearheads of class I, one of class II, and one halberd. In addition, there is the recently discovered grooved dagger, found with a perforated stone axe-hammer of

Snowshill type accompanying a contracted skeleton under a round barrow at Chippenham—the only metal object of this period in the County with any associations. This find is an outlier from the recently distinguished Early Bronze Age culture of Wessex.

The scarcity of the earliest metal objects is doubtless due to the over-running of the County by the Beaker people, who first reached here in a more or less "neolithic" stage of culture. Of the two main groups distinguished in south-eastern Britain, Cambridgeshire was affected mainly by the "A" beakers, the "B" group being represented only by a solitary example from Isleham Fen. The "A" beakers are found distributed round the entire margin of the Fenland, and evidently the Wash formed a main avenue of entry for these people. In Cambridgeshire, they settled upon the larger fen islands (March, Doddington, Ely), and on the low-lying Burnt, Burwell, Isleham, Lode, and Quy Fens. They also pushed up the valleys of the Snail, the Cam, and the Granta; and finds from Therfield Heath and Hitchin suggest that some of them pressed down the Icknield Way to Wessex. The Beaker pottery from the County is outstandingly rich, and special mention should be made of the three handled beakers; two of them are of the rare straight-sided type.

The open settlement sites, found on sandy hillocks at Shippea Hill (Plantation and Peacock's Farms) and Isleham, have produced quantities of sherds from beakers with rusticated surface; and, in addition, they have yielded cord-impressed sherds of "food-vessel" affinities with internally bevelled rims. The latter are highly significant, for they show that the "native" element was not submerged by the Beaker invaders. It was among a mixed population that the metal types of the full Early Bronze Age circulated. The flint work from the fen sites is of a high standard; shallow pressure flaking is seen to great advantage on the scrapers, the barbed and tanged arrowheads, and the plano-convex knives.

The County is rich in contracted inhumation burials accompanied by beakers, the graves in every case being flat. Generally, the graves occur singly, but an undoubted cemetery was destroyed over a period of years in a sand-pit off Springhead Lane, Ely, where the discoveries included at least 14 human jaws, an "A" beaker found together with a perforated stone axe-hammer near the head of a skeleton, and a "C" beaker in company with another skeleton. Among the objects associated with isolated beaker burials may be mentioned "a bull's horn", found with the Wilburton Fen beaker in 1847; while a grave group, recently discovered at Little Downham, yielded an "A" beaker, a flint dagger,[1] a flint knife, a V-perforated button, and a pulley ring of shale.

[1] Seventeen similar ones have been found loose in the County.

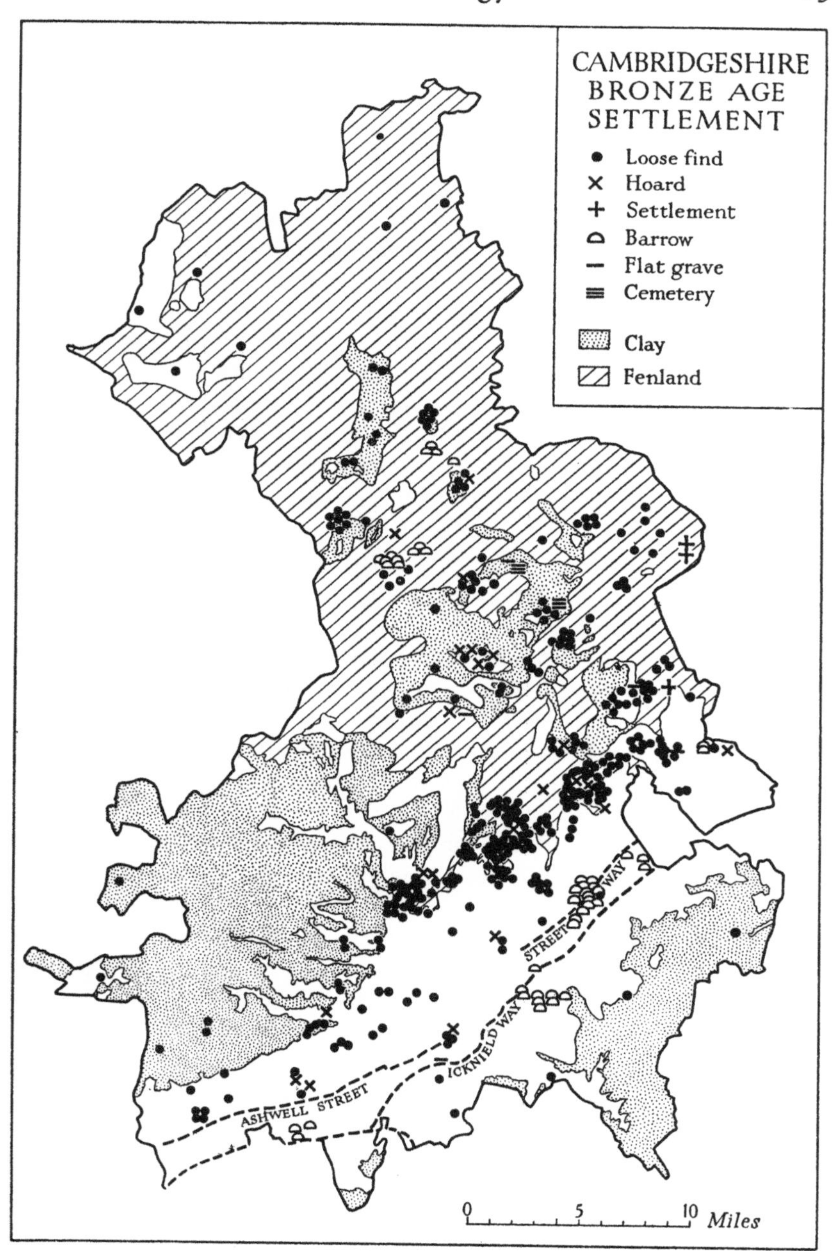
CAMBRIDGESHIRE
BRONZE AGE
SETTLEMENT
Loose find
Hoard
Settlement
Barrow
Flat grave
Cemetery
Clay
Fenland
ASHWELL STREET
ICKNIELD WAY
STREET
WAY
0
5
10
Miles

Fig. 18.

*Middle Bronze Age.* The Middle Bronze Age in Cambridgeshire was a period of prosperity undisturbed by invasion. Bronze implements came more widely into use; no fewer than 122 palstaves, two daggers and dirks, 20 rapiers, and 27 looped spearheads, have been recorded from the County. Irish gold also found its way into the area in some profusion; and there have been two famous finds from Grunty Fen, Wilburton, one in 1844 comprising a splendid multiple-ribbon twisted gold torc with solid terminals and three looped palstaves, and another in 1850, which included a similar torc, a part of a bronze rapier, and a gold bracelet with attached ring money.

The pottery in use locally at this time was the overhanging-rim ware, known from a single open settlement in Isleham Fen, and from many burials. Cremation was the dominant rite, the ashes being contained either in an urn or in some kind of bag. It is to this period that many of the Bronze Age round barrows in the County belong; although some had already been erected in the Early Bronze Age, the cremations being inserted secondarily. The barrows on the chalk belt are strung out along the line of the Icknield Way in five main groups; at Chippenham, on Newmarket Heath, at Upper Hare Park, near the junction of the Fleam Dyke and the Icknield Way, and on the downs east of Royston in the parish of Melbourn. The group of fen barrows, found mainly within the triangle Mepal-Manea-Chatteris, is significant from its occurrence at heights barely above mean sea-level; this emphasises the geographical conditions prevailing in the earlier stages of the Bronze Age in this region.

*Late Bronze Age.* The Late Bronze Age saw a further substantial increase in the use of metal. Finds are nearly three times as numerous as in the preceding period and they embrace a much wider range of types. A high proportion of the bronzes (some 374 out of a total of 495) comes from 19 hoards. Some of these, like the pair of shields from Coveney Fen, are probably "votive"; others, e.g. the leaf-shaped swords from Chippenham, mark a local metallurgical industry; but most of them belong to the classes known as merchants' hoards (e.g. the Wilburton hoard of 163 pieces—mostly spearheads), and founders' hoards (e.g. the hoard at Green End Road, Cambridge, containing many broken objects and over 17 lb. of metal cakes). Together, these reflect the extensive trade responsible for the introduction of a flood of exotic types, mainly of Central European origin, to Cambridgeshire. It is evident, from the fact that all the leaf-shaped swords belong to the "V" type, that the "U" sword complex did not affect Cambridgeshire in its earliest stage. Further, although marginal to an area strongly affected by the "Carp's tongue" sword complex, Cambridgeshire was hardly influenced; none of the characteristic swords has been found in the County, and only one fragmentary winged axe.

Although exotic influences resulted in a revolution of the metallurgical industry at this time, it is likely that the change was mainly brought about by trade rather than by ethnic invasion, since the County is strictly marginal to the area of the so-called Deverel-Rimbury pottery. Finds of pottery of this class consist only (1) of a few sherds from a circular trench (2–3 ft. deep with a diameter of 68 ft.) at Swaffham Bulbeck, (2) the lower part of a finger-printed barrel urn from Chesterton, and (3) the upper part of a small pot with slashed rim and applied bosses from the Little Thetford–Fordham causeway. Evidence that the Middle Bronze Age overhanging-rim urn pottery in this part of the country continued into the succeeding period is supplied by the material from a settlement site in Mildenhall Fen, only just over the Suffolk boundary, where a fusion between the two wares can be detected. It is thus probable that some of the "Middle Bronze Age" burials from the County really belong to this period.

## THE EARLY IRON AGE

The material available for the study of the Early Iron Age in Cambridgeshire is scanty, and only an insignificant proportion has been obtained from scientific excavation. This is largely due to the scarcity of "hill-forts", or other prominent sites, that might have invited excavation. There are no certainly established Early Iron Age defended sites in the County, apart from (1) Wandlebury, a circular triple-banked site with a diameter of about 1000 ft. crowning the crest of the Gogmagog Hills, and (2) the War Ditches, a smaller single-ramparted site on a spur of the same hills. The slight excavations carried out at the War Ditches, prior to the partial destruction of the site, prove that the ditch was quarried by people of Early Iron Age "A" culture, although the surviving material is meagre. The key to the interrelations of the Early Iron Age cultures of Cambridgeshire must be sought in further digging in the surviving portions of the War Ditches and in Wandlebury, a site that encloses a private house, and which has yet to be excavated.

The normal settlement was open and undefended, generally without surface indication, but sometimes delimited by a low bank as at Bellus Hill, Abington Pigotts. It is perhaps for this reason that little systematic work has been done, and that the discovery of settlement sites of this period has invariably been accidental. From the meagre information available it would appear (Fig. 19) that settlement was concentrated in the valleys of the Cam above Cambridge, the sites commonly being placed in pairs either side of a ford, e.g. Grantchester and Trumpington, Barrington and Foxton. The clay areas remain completely blank apart from stray finds of Belgic coins, many of which doubtless continued to circulate at a later date, and, in any case, can hardly be regarded as indicative of settlement. The evacuation of the Fenland, already begun in the later stages of the Bronze Age, was virtually complete by the Early Iron Age, with the exception of certain of the larger islands. Coin finds in the north of the County may well relate to the Romano-British settlement of that region. The only finds from the fens between Cambridge and Isleham consist of a stray brooch and a discarded chariot-wheel boss of Belgic type. Yet, in the Bronze Age, this area was the most populous district in Cambridgeshire.

The County was affected mainly by two successive spreads: the so-called "A" culture, and the "C" (or Belgic) culture. If the number of brooches and pins of early La Tène type is any criterion, it would appear that the "A" culture spread into the County not later than the latter half of the fifth century B.C. The distribution of the finger-impressed pottery, in Cambridgeshire and neighbouring counties, certainly suggests the Wash as the main entrance, although certain elements, such as the "plugged-in" handle, may well have come here from Wessex by way of the Icknield Way and kindred routes.

The "B" culture is represented by the famous Newnham Croft burial, at Cambridge, accompanied by outstandingly rich grave goods; but settlement material is entirely lacking.

Although the "C" culture was essentially intrusive, penetrating from the south about the middle of the first century B.C., it seems unlikely that there was any complete break in the continuity of the Early Iron Age settlement of the County. Many sites, such as Abington Pigotts and Hauxton, have yielded pottery from both cultures. Among the numerous cremation cemeteries of the County, only one—that at Guilden Morden—has been scientifically excavated, and comparatively few graves in this were of pre-Roman age. An iron fire-dog, with ox-head terminals, from Lord's Bridge may have come from the vault of some important individual, but little is known of the circumstances of the find.

There is ample evidence for trade at this time. Pottery from the kilns of Arezzo in Italy was imported through Gaul, while several finds of amphorae of Mediterranean type probably indicate trade in wine or oil. Such imports were doubtless paid for in part by the export of slaves; a fine slave chain with six collars from Lord's Bridge may be a reminder.

The native coins, in which Cambridgeshire is rich, show that most of the County fell within the territory of the Catuvellauni, whose princes[1]

[1] Tasciovanus (20/15 B.C.–A.D. 10) and Cunobelinus (A.D. 10–40/3).

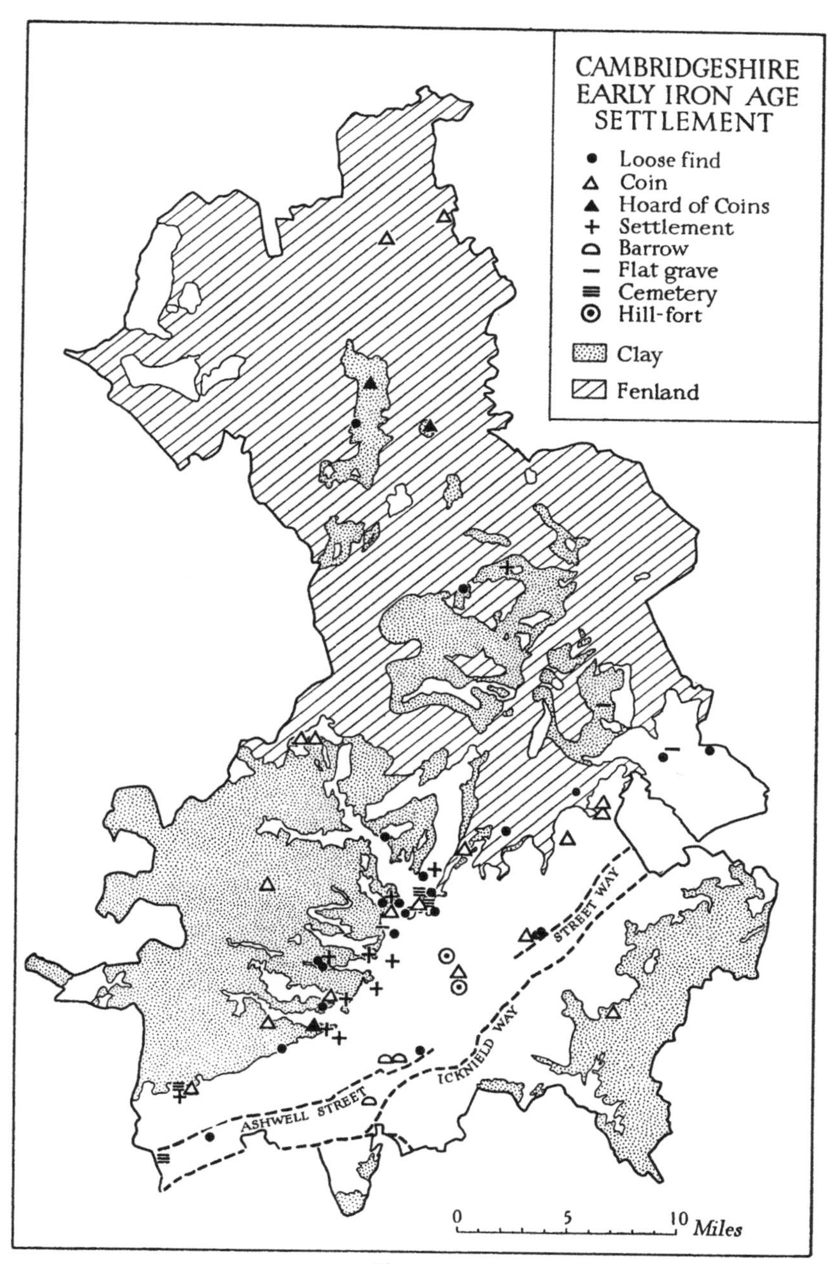
CAMBRIDGESHIRE
EARLY IRON AGE
SETTLEMENT
Loose find
Coin
Hoard of Coins
Settlement
Barrow
Flat grave
Cemetery
Hill-fort
Clay
Fenland
STREET WAY
ICKNIELD WAY
ASHWELL STREET
0
5
10 Miles

Fig. 19.

minted their coinage at Verulam and Colchester during the sixty years prior to the Roman Conquest. But the northern part of the County, and a fringe to the east of the Devil's Ditch, formed part of the tribal area of the Iceni, hoards of whose coins are known from March and Wimblington.

## ROMANO-BRITISH TIMES[1]

In Roman, as in prehistoric times, the human settlement of Cambridgeshire was dominated by its physical features and its superficial geology. With the exception of the silt-land farms, the distribution of Romano-British population was not fundamentally different from what it had been in the Early Iron Age, and it is clear, despite the existence of some finds on the Boulder Clay in the south-west of the County, that no serious attack was made on the considerable areas of scrub which must have covered much of the district (see Fig. 20).

One of the dominating features of life in all the Roman provinces was the presence of a developed road system and, in greater or lesser measure, of organised town life. The dominant feature of the Roman road system in this district is the Ermine Street, entering the County at Royston, passing out of it towards Huntingdon at Papworth Everard, and taking its name from the Cambridgeshire hundred of Armingford.

Secondary roads converged on Cambridge. From the south-east, came the *Via Devana* from Colchester,[2] which is probably the earliest Roman road in the region. Entering Cambridge from the south-west, was the Akeman Street which branched from Ermine Street, north of the Cam crossing, to continue its course north-east to Ely, and possibly to Littleport. There was also a local road from Braughing, through Great Chesterford, to join the Icknield Way at or near Worstead Lodge on the line of the so-called *Via Devana*. In the west of the County there was also the secondary road from Sandy to Godmanchester which now forms the County boundary for a short way. Last comes the Icknield Way, which must have continued in use in Roman times, though there is no evidence that it was metalled or otherwise regulated by Roman standards.

During earlier times, the clay areas of the upland had been almost without population, which seems to have been concentrated partly along the chalk belt and partly in the valley of the Cam. It might be supposed that the Roman road system, cutting through these clay areas both in the

[1] By C. W. Phillips, M.A.

[2] According to Fox, this originally may have missed Cambridge to join the Ermine Street at or near Caxton, but was later re-aligned to pass through the town and lead north-west to Godmanchester. This view, however, seems less likely now than in 1923.

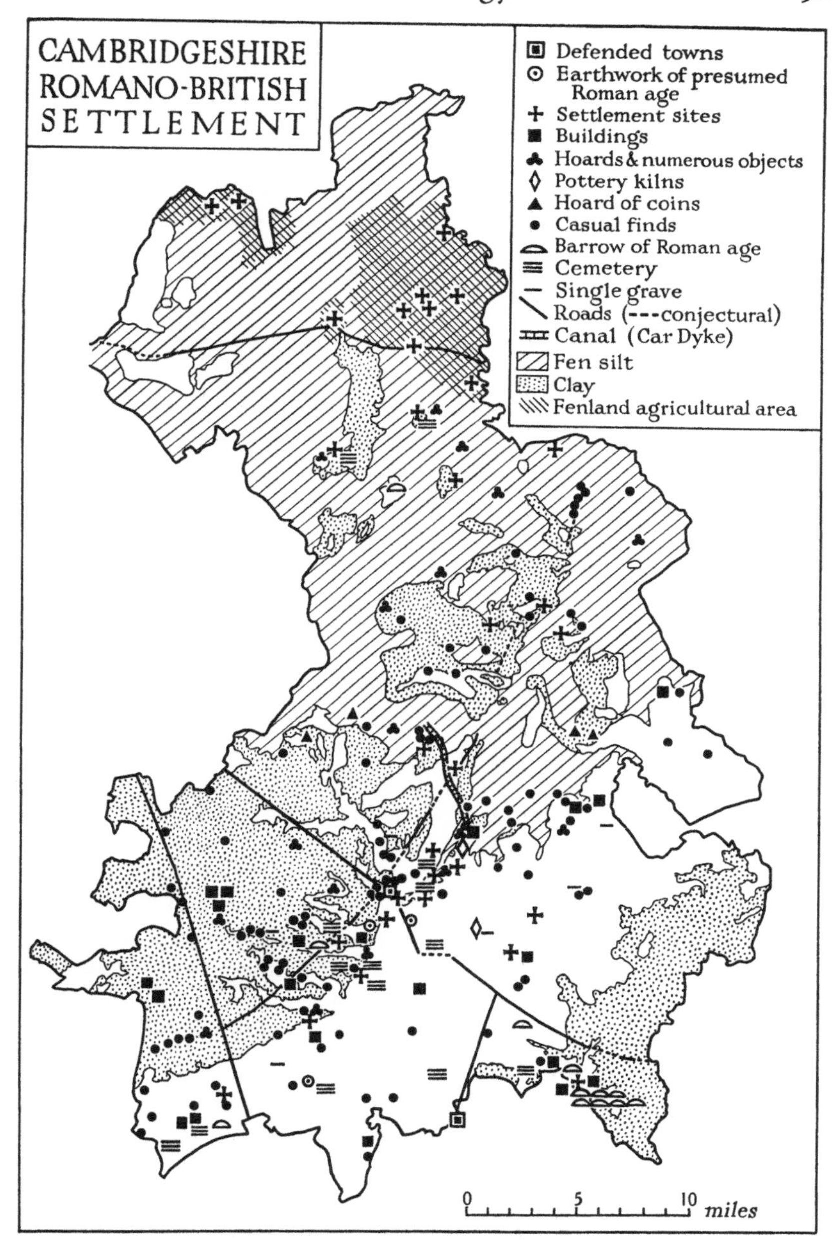


Fig. 20.
The Car Dyke joined the Old West River (not marked here) at Lockspit Hall. See Fig. 7.

south-east and south-west, would have induced settlement, but there is little evidence of this. Penetration of settlement into the uplands was still by way of river valleys; those of the Linton Granta, the Newport Cam, and the Bourn Brook, show this very clearly, while the courses of the roads are almost without settlement. In thrusting their way into the clay lands by the river valleys the Romano-Britons were carrying on, in larger numbers and with better equipment, a movement that had begun in the Early Iron Age, but they do not appear to have made any serious effort to occupy the wooded country as a whole. This task was reserved for the Anglo-Saxons.

The general style of rural life seems to have been humble. No country house of any importance has yet been found anywhere in the County. There are no indications of any industrial activity except for a pottery for coarse wares at Horningsea which enjoyed a fairly wide local market. Few individual finds of much importance have been made in the County, but the Fenland has yielded a number of good pieces of Roman pewter, and there is also the remarkable group of cult objects associated with the worship of the Emperor Commodus as Romanus Hercules found in Willingham Fen. The County, too, has some notable examples of the high conical type of barrow of Roman age. The Bartlow group, though badly damaged, remains the finest of its kind in Britain.

In the Fenland, an extensive Romano-British occupation has been recently demonstrated, more particularly on the silt areas and on certain islands. This settlement was agricultural, and the region of maximum farming activity seems to have been to the north upon the siltlands of south-eastern Lincolnshire.[1] Before this discovery, the frequency of stray Romano-British finds in the fens had been a puzzling fact.

The exploitation of the silt areas began at once after the Roman conquest, and a large population of relatively well-to-do peasant cultivators spread over a region which, it has been suggested, was administered as a domain of the Roman people, though this fact can only be inferred from the general conditions, and does not rest on any confirmatory discoveries. In the less favourable parts of the Fenland, there was a fair sprinkling of folk living in small groups. Many of their sites had close relation to watercourses, but both their house sites, and the adjacent small fields, were carefully protected against tidal floods, for it must be understood that, at this time, tides came far up the wide fen estuaries. Towards the close of the third century, conditions seem to have deteriorated. Whereas it may normally have been unnecessary to organise any drainage works, a slight subsidence of the whole fenland basin may have made the last one and a half centuries

[1] See p. 20 above.

a time of increasing difficulty for the fenland cultivators. Alternatively, it is possible that the disaster may have been due to a combination of tide and wind causing a general breach in the natural silt defences which the sea built against itself around the southern margins of the Wash. In any case, matters had reached such a condition by the fifth century, that the general abandonment of the region was due to take place whether the Anglo-Saxons had come or not. It is significant that the latter made no attempt to settle anywhere in the fenland basin, and that they confined themselves to the country round the edge. In view of their farming habits it is unlikely that they would have failed to occupy a region that had been intensively and successfully cultivated, if it still remained in any physical condition favourable to their enterprise. The Saxons, for the first time, subdued and occupied the scrub-clad uplands of the County, but they had only been able to make a sparse settlement in favourable parts of the Fenland by the time of the Domesday Survey, half a millennium after their first settlement.

There are no large urban sites in Cambridgeshire. Roman Cambridge was a subrectangular area about 26 acres in extent defended by a bank and ditch of late date. It was a road junction of local importance, but no architectural remains of any kind have ever been found in its area. We are compelled to envisage little more than a village built of wood, clay, and thatch.

Conditions have not been favourable for finding out much about Roman Cambridge because at various times a great thickness of top soil has been removed from one of the most hopeful areas, but, as a result of the finds made recently during the building of the new Shire Hall, it can now be said that there was some occupation of the site in the middle of the first century A.D. and that the former existence of a military camp belonging to the period of the Claudian conquest is probable. No trace of a wall has ever been found round Roman Cambridge, though Bowtell, in the early nineteenth century, reported that some traces were then visible in his judgment close to the Huntingdon Road's exit from the enceinte.[1]

The only other Roman town in the district was just over the Essex border, at Great Chesterford. This was a more important centre with a strong wall, much of which was still visible in Stukeley's time, though all above ground has now vanished. The numerous and important finds made here at different times suggest that it was an active local centre of the smaller kind.

[1] T. Bowtell, MSS. in Downing College Library, Cambridge.

## THE ANGLO-SAXON PERIOD[1]

The Anglo-Saxon Age in eastern England can be divided into three periods:

(1) The period of the Pagan Cemeteries which may be thought to include the fifth, sixth and part of the seventh centuries. This was the age of early settlement, and may perhaps be compared with the seventeenth and eighteenth centuries in North America.

(2) The Early Christian Period, which of course overlaps the Pagan Period to some extent. It included part of the seventh, the eighth and part of the ninth centuries.

(3) The Viking Age, which closed with the final extinction of Anglo-Danish culture towards the end of the eleventh century.

These periods are by no means watertight compartments, and, compared with the Early Iron Age, the Anglo-Saxon period is very imperfectly understood. It is seldom realised that the Anglo-Saxon period lasted for nearly seven hundred years, and that, except for the Pagan Cemeteries, it is extremely difficult to locate sites whose excavation would throw any light on conditions of those times.

In the Cambridge area, recent years have seen a great advance in the study of the period. There have been excavations spread over several years upon the big linear earthworks that are so outstandingly a feature of the County. The following earthworks are shown in Fig. 21, starting with the most north-easterly: Devil's Dyke, Fleam Dyke (in two parts), Brent or Pampisford Ditch, Bran or Heydon Ditch, and the Mile Ditches. Two cemeteries have been discovered which belonged to the period of overlap between the pagan and Christian periods. Small villages of both pagan and Christian periods have been found and investigated either within or not far beyond the boundaries of the County. Finally, several small excavations have provided a very hopeful start with the study of the pottery belonging to the last three centuries of the period. But, with the exception of linear earthworks, the student of Anglo-Saxon England is very much handicapped by having to wait till a chance find may give him a clue upon which to work. There are no villages surrounded by great earth-ramparts that are so helpful in the study of the Early Iron Age in some parts of Britain. The great majority of Saxon villages are beneath those of the present day, and are therefore irretrievably lost to Archaeology. Manor-sites may perhaps offer a slightly better field for study, but there again the chance of finding one not occupied by later buildings is remote.

[1] By T. C. Lethbridge, M.A.

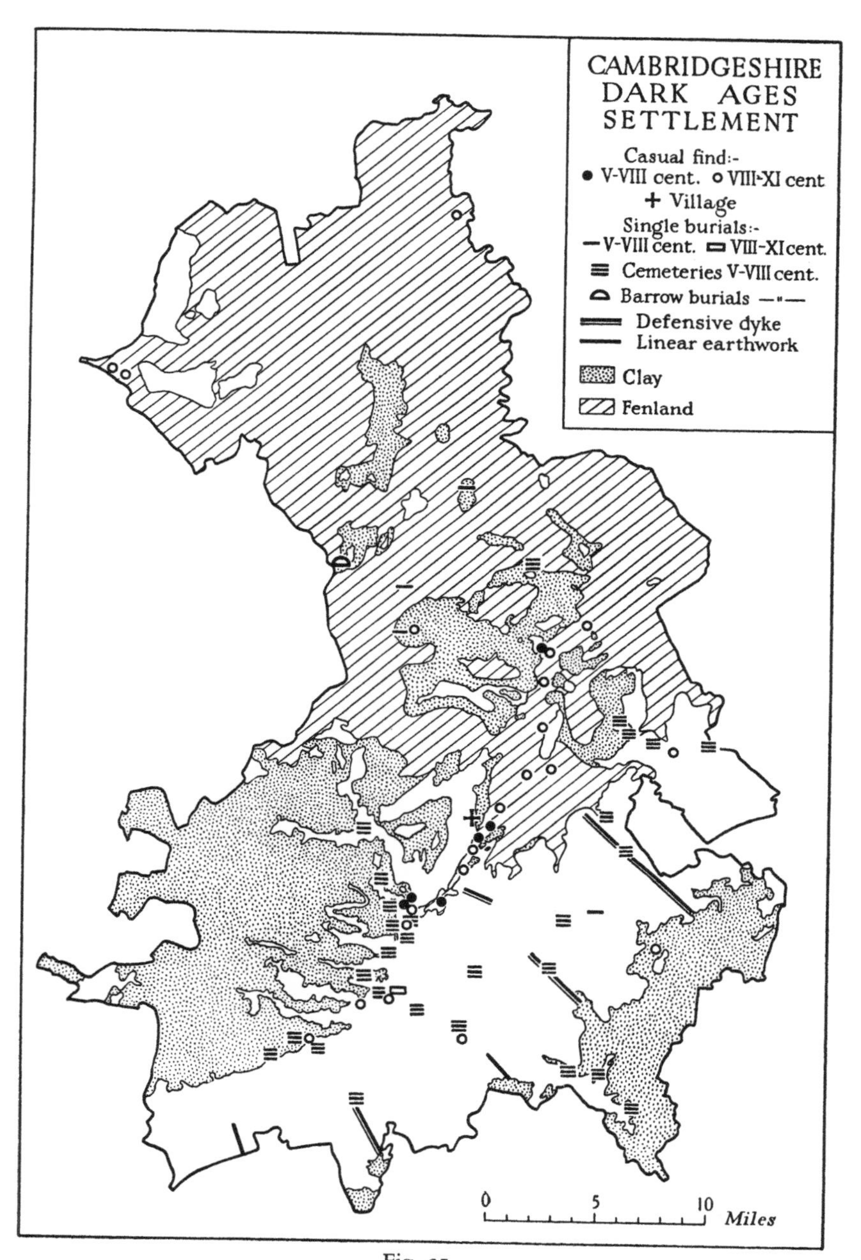
CAMBRIDGESHIRE
DARK AGES
SETTLEMENT
Casual find:-
• V-VIII cent. ○ VIII-XI cent
+ Village
Single burials:-
— V-VIII cent. ▭ VIII-XI cent.
≡ Cemeteries V-VIII cent.
Barrow burials —"—
Defensive dyke
Linear earthwork
Clay
Fenland
0
5
10
Miles

Fig. 21.

(1) *The Pagan Period.* The evidence for this period in Cambridgeshire is abundant. More than two dozen burial sites are known. Most of these have produced a considerable number of graves (sometimes running into hundreds), and most of the burials were accompanied by grave-goods. There is therefore a very large collection of objects from these finds on view in the University Museum of Archaeology and Ethnology.

The men were usually buried with their weapons and often with food for the next world; while the women were dressed in woollen clothes with numerous ornaments such as brooches on the breast, beads in long festoons round the neck, girdle-hangers at the waist, and clasps at the wrists. In some cemeteries (e.g. at Little Wilbraham) numerous cremation burials have been found, but, on the whole, cremation seems to be the exception rather than the rule in Cambridgeshire. Fig. 21 shows that the cemeteries are confined to the fen margins and to the river valleys. No cemeteries have, as yet, been found on the Boulder Clay covered uplands. This raises several points which are difficult to explain, for although the clay areas were apparently devoid of pagan Saxon settlement, yet the Domesday Survey shows them to be as densely occupied as the river valleys (see Fig. 22). It is not clear whether this means that the uplands were unpopulated in pagan times, or whether, perhaps, survivors of the Romano-British population lived on them. Another phenomenon, as yet unexplained, is the scarcity of settlement along the River Ouse. Cemeteries are so numerous in other Cambridgeshire valleys, and in those of the adjacent parts of Norfolk and Suffolk, that one would have expected a similar concentration along the Ouse. There are some burial sites along the river, but they contain very few burials until Kempston is reached near Bedford.

Only one village of the Pagan Age has as yet been explored in Cambridgeshire. This is situated on the east bank of Car Dyke at Waterbeach, and its rubbish overlay the silting of this Roman canal. The huts were of the same semi-pit-dwelling type that has been noticed elsewhere. A village of the Viking Age at St Neots just outside the County had huts of similar form. It has not been definitely established whether these pit-like hovels were really the living rooms of the period, or whether they were undercrofts to upper stories as indicated in the Bayeux Tapestry.

Of the great linear earthworks of the County, the Devil's Dyke is one of the most spectacular monuments of its kind in Britain,[1] while the Fleam Dyke is little less remarkable. Not much can be seen now of the Bran

[1] The Devil's Dyke is 7½ miles long with a rampart 15 ft. high. It runs across the open chalk country from the Fenland to the one-time wooded clayland, and has been laid out in three straight sections, the north-westerly one of which was apparently aligned on a Roman canal or lode which ran from Reach to Upware.

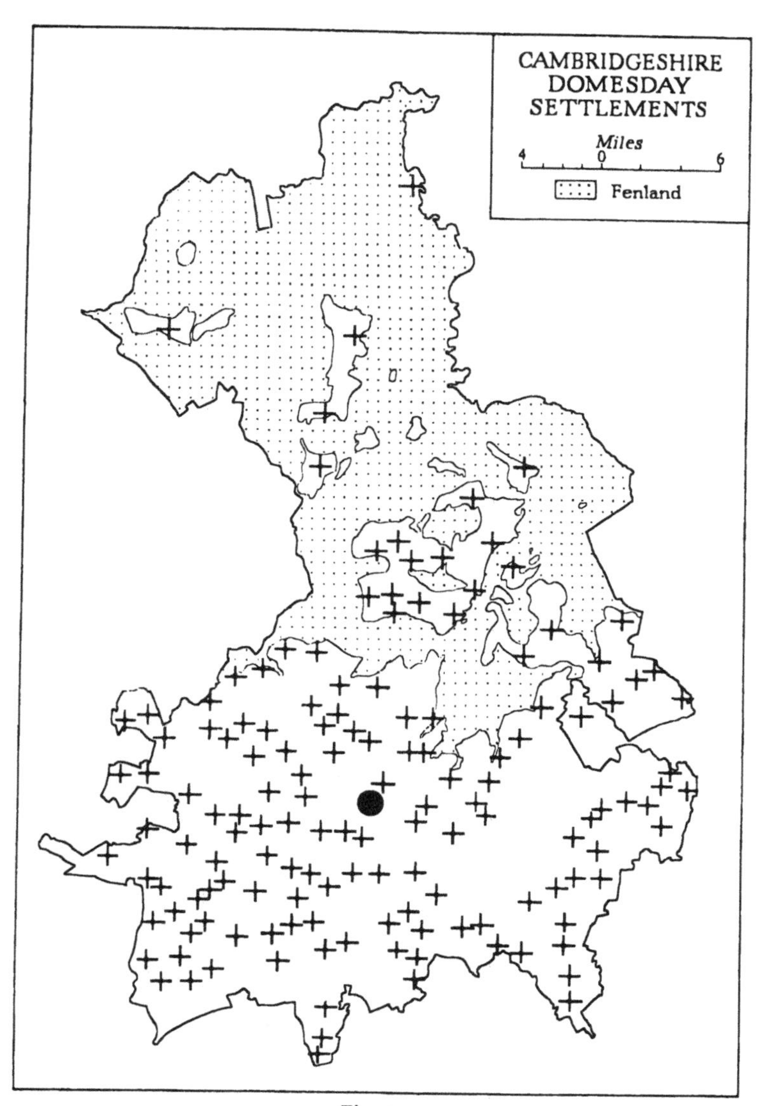


Fig. 22.

From H. C. Darby, "The Domesday Geography of Cambridgeshire", *Proc. Camb. Antiq. Soc.* xxxvi, 39 (1936). The town of Cambridge is indicated by a black circle. Compare the distribution of villages on the upland with the empty clay areas of Fig. 21.

Ditch, while Brent Ditch has no bank. All four appear to represent attempts to hinder communication along the open chalk downland during the Pagan Age. Burials of warriors with their weapons have been found at Devil's Dyke and Fleam Dyke, beside the Worsted Street. Possibly there were also burials in the Brent Ditch. At the Bran Ditch, groups of decapitated skeletons have been found at two localities. They had associated objects of this period. Excavations[1] have shown that these linear earthworks were, in all probability, constructed during the early wars of the Heptarchy, when Penda of Mercia (*c.* 655) overthrew the rulers of East Anglia.

Other indications of warfare in early settlement times may be deduced by finds of early swords, spears, shields, and human bones in the Cam at Clayhithe, opposite the end of the Car Dyke.

(2) *The Early Christian Period.* Generally speaking, this period has left but the scantiest of material remains in the County. A few chance finds may be seen in the University Museum. But in recent years extensive cemeteries at Burwell and Shudy Camps have been explored. These appear to belong to the period of overlap, and may perhaps have continued into the eighth century. It is probable that another cemetery exists at Foxton, while a burial at Allington Hill may perhaps be that of an important man killed at the Dykes in the seventh century.

(3) *The Viking Age.* Finds of small objects, weapons, and human skeletons, at Hauxton Mill probably indicate trouble there about the year 870. Pottery of the Viking Age is now being recognised from many localities, especially in the town of Cambridge itself, and, to a less extent, down the Ouse Valley, while small crosses and grave slabs characteristic of the district have been widely noted.

But the chief finds relating to this period are the rich series of weapons found in the rivers. For the most part, these may belong to the final campaign when William I overcame the last resistance of Hereward the Wake and his followers in the Isle of Ely. By this time, the Domesday book (see Fig. 22) presents a complete picture of settlement in the County, and sums up the economic activity of the Anglo-Saxon period.

[1] Objects of the later part of the Roman period were found beneath the Fleam Dyke and Devil's Dyke, while skeletons of the Anglo-Saxon period and late Roman pottery were found under the vallum of the Bran Ditch.

CHAPTER SEVEN

# THE PLACE-NAMES OF CAMBRIDGESHIRE[1]

By P. H. Reaney, LITT.D., PH.D.

CAMBRIDGE, GRANTCHESTER, AND ELY WERE RIGHTLY derived some thirty years ago by the great Cambridge philologist and pioneer of English place-name studies, the late Prof. Skeat.[2] A few scholars in other fields hesitated to accept his etymology of Cambridge but recent advances in the study serve only to confirm it. The earliest reference to the town is Bede's *Grantaceastir* (*c.* 730), "the Roman fort on the Granta". This would normally become "Grantchester", but the reference is undoubtedly to Cambridge and not to the modern Grantchester which appears in early sources as *Granteseta*, "the settlers on the Granta". As early as 745, in Felix's *Life of St Guthlac*, had come the change in the second element which has given rise to the present-day name of Cambridge (*Grontabricc*). The site of a Norman castle and a centre of Norman administration, the town was subject to strong Norman influence which had its effect on the name, until, through such forms as *Cantebruge* (*c.* 1125), *Cauntebrig'* (1230) and *Caumbrigg* (1353), the ancestor of the modern spelling was reached in *Cambrigge* (1436).

Grantchester is an interesting example of phonetic change and popular etymology resulting, ultimately, in the form that Cambridge should have had. *Grantsete* became *Gransete* and *Grancete*, pronunciations which suggested an analogy with such names as Leicester and Worcester. The name was accordingly spelled *Granceste*, *Grancestre*, *Granceter*, and finally *Granchester*, a spelling which has not yet been noted earlier than the seventeenth century.

The river on which Cambridge stands is known in various parts of its course as Granta, Cam, and Rhee. Granta, the real name, is unique and pre-English, meaning, probably, "fen river" or "muddy river". When Cambridge came to be known as *Cantebrigge*, this was interpreted as "the bridge over the *Cante*", an artificial back-formation found from 1340 onwards. Similarly, the modern *Cam* is a later back-formation from the

[1] This essay is based on a preliminary survey of material so far collected for a volume on "The Place-names of Cambridgeshire" to be published by the English Place-name Society. The discovery of further material may necessitate some modification of detail.

[2] W. W. Skeat, *The Place-Names of Cambridgeshire* (Camb. Antiq. Soc. 1901).

spelling *Cambrigge*, and this was sometimes Latinised as *Camus*. *Rhee* is from Old English *ēa* "river"; *æt þǣre ēa* "by the river" became Middle English *at ther ee*, which was wrongly divided as *at the ree*. In 1285, William *atte Ree* lived by the Granta at Grantchester.

Ely occurs first, in Bede's *Ecclesiastical History*, as *Elge* "eel-district" The second element is the archaic *ge*, corresponding to the German *gau*, found also in the names Surrey, Eastry, Lyminge, and Sturry in Kent, and Vange in Essex. Here, too, popular etymology was early at work and, already in the Anglo-Saxon version of Bede, the name appears as *Elig* "eel-island". Domesday Book records the yield of innumerable eels from the fisheries of Ely and renders of eels were common elsewhere in the island. In Sutton, too, was a place called *Cappelode*, a name identical with the Lincolnshire Whaplode, "eelpout stream".

These names are of especial interest because of their age. Cambridge and Grantchester contain the Celtic name of the river Granta. Grantchester, too, was a folk-name—"the settlers on the Granta", the second element being that found in the names of such large districts as Dorset and Somerset. Ely was the name of the whole island, called by Bede a *regio* and in the Anglo-Saxon version *þēodlond*. These names are of high antiquity and may well date from the Anglian settlement which archaeologists agree in placing in the latter half of the fifth century.

From the earliest periods, the estuary of the Wash has been a magnet for successive hordes of invaders from beyond the sea, and its river valleys have afforded an easy way inland. The Angles found the Fenland largely unattractive for settlement. They pushed on, and clear evidence of their former presence has been found in a number of cemeteries in the Cam Valley, where the finds point clearly to a settlement by the end of the fifth century.[1] Numerous place-names of recognised early type might therefore be expected here. But of the oldest type, that ending in *-ingas*, there is only one, Kirtling, as against some twenty-four in the neighbouring county of Essex, where, too, names pointing to heathen worship are common. Of these, in Cambridgeshire there is not one. The probable explanation of this curious contradiction—the absence of numbers of place-names of high antiquity combined with clear archaeological proof of very early settlement—is that the struggles for supremacy between East and Middle Anglian and the Danish invasions resulted in such confusion, devastation, and depopulation that memory of the names of all but the most important places was utterly lost.

Names of early type do exist. Badlingham, Cottenham, Dullingham,

[1] R. G. Collingwood and J. N. L. Ayres, *Roman Britain and the English Settlements* (1936), pp. 386–7.

and Willingham, "homes of the followers of Bæddel, Cotta, Dulla, and Wifel", are of a very early type. Haslingfield, "the open country of the dwellers by the hazel-wood", is an early name and the site of a fifth-century cemetery. Armingford Hundred and Arrington, "the ford and the farm of the people of Earna", to whom Ermine Street also owes its name, are early formations, and the occurrence of their name in three distinct places suggests that the *Earningas* were of some importance.

The element *ham* is, on the whole, earlier than *tun*, but some *-ton*-names are older than some in *-ham*. Sawston, earlier *Salsingetune*, "the farm of the followers of Salsa" (a personal-name otherwise unknown in England), and Hinxton, deriving from *Hengestingatun*, "the farm of the followers of Hengest" (a name known to have been borne by one of the earliest of the invaders of Kent), are undoubtedly of greater age than such names as Fordham, Coldham, and Downham; whilst from their very meaning the three examples of Newnham and the two of Newton must be comparatively late. So, too, are place-names which provide evidence for women as landholders: Wilbraham (the site of a cremation cemetery) and Wilburton, "the *ham* and *tun* of Wilburg" and Babraham, "the *ham* of Beaduburg".

Although Cambridgeshire was part of the Danelaw, it was never so thoroughly Scandinavianised as Norfolk and Lincolnshire. Caxton and Croxton contain the Scandinavian personal-names *Kakkr* and *Krókr* respectively.[1] Conington is a Scandinavianising of an English *Kington*, and Carlton of *Charlton*, "the farm of the ceorls". "Toft" is a Scandinavian word meaning "the site of a house and its buildings", or "homestead". It survives as the name of a parish, and formerly occurred as a field-name in eight other parishes. But, in general, the parish-names of the County give no such clear evidence of Danish influence as do those of Lincolnshire.

Minor names, however, suggest that Danish influence was not negligible, and that, in places, it was strong. Denny, "the island of the Danes", implies that Danes were not numerous in the neighbourhood. Such words as *holm* "small island or dry land in a fen" (a common field-name), and *bigging* "building" (e.g. Biggin Abbey in Fen Ditton), while Scandinavian in origin, are no criterion of Scandinavian settlement. But Clipsall in Soham, together with three field-names, contain the Scandinavian personal-name *Klippr*, and Hoback Farm is a hybrid, "the beck or stream in the hollow". On the borders of Huntingdonshire there are two examples of *lundr* "a grove", and several of *krókr* "a bend". But more interesting is the occurrence of the Scandinavian *kirk* and its interchange

[1] They (and Toft below) may well point to a late settlement by Scandinavians in this wooded area. For the woodland see p. 52 above.

with the English *church*. In Tydd St Giles, both Kirkgate and Church Lane occur, and in Thriplow to-day there is a Church Street where in the thirteenth century a neighbouring field was *Kirkefeld*. So, too, in Newton-in-the-Isle, the modern Church Croft is paralleled by the seventeenth-century *Kyrkelandefield*, and other lost field-names give further examples of Scandinavian influence. A similar interchange occurs in the second element of Landwade. The modern spelling represents the English *wæd* "ford" but early forms often have the Scandinavian *vað*.

In the north of the County, on the Lincolnshire border, there are such names as Gate End Bridge (Parson Drove), Eaugate Field, Fengate Field, Kirkgate, and Newgate (all in Tydd St Giles). These contain the Scandinavian *gata* "road", common in such street-names as Waingate in Sheffield, Briggate in Leeds, etc. It is often impossible to distinguish this from the common word *gate*. Kirkgate, for example, might well mean "church-gate", but Eaugate and Fengate can hardly mean "gate leading to the river or the fen", nor does the common *gate* make much sense in Gate End Bridge. Clear proof, however, is forthcoming from *Kyrkestrete* (1393) in Leverington described in 1486 as "regiam viam vocatam *le Kirkgate*", and from a "highway called *Crossegate*" in 1438 at Tydd St Giles.

In the Isle of Ely there is no parish-name of Scandinavian origin. The evidence suggests the naming of minor places in a more or less settled time, and the substitution of English terms by similar, corresponding Scandinavian ones, some of which were common to various districts where the settlement was not strong.

French influence is much less in evidence. Marmont, the name of a priory in the fens in Upwell, was transferred from France, *mirum montem* "the famous hill". Not far away, in Elm, is Beauford, "the fair ford", whilst in Willingham an old earthwork, Belsars Hill, is identical in origin with the Durham Bellasis and the Belsize of Hertfordshire and Northamptonshire; these are derived from *bel assis* "the fair seat", and the name thus loses its historical associations with the fictitious Belasius, the knight who, in the campaign against Hereward, acted as the Conqueror's Commander-in-Chief.[1]

Apart from these names, French influence is confined to the commemoration of the Norman families of *Everard* de Beche in Papworth Everard; *Agnes* de Papworth in the erroneously canonised Papworth St Agnes; the Colvilles in Weston Colville; the Bolebecs in Swaffham Bulbeck; and in such manorial names as De Fréville Farm in Great Shelford, D'ovesdale Manor Farm in Litlington, Lacy's Farm in Duxford, and Lacies Farm in Grantchester (from the Lacis, earls of Lincoln).

[1] E. Conybeare, *Highways and Byways in Cambridge and Ely* (1910), pp. 283, 292.

Southern Cambridgeshire consists of three well-defined areas, two belts of clay on east and west, with an intervening stretch of chalk (see Fig. 29). The claylands were presumably wooded in early times, but the chalk was always open country, along which ran the Icknield Way (see Fig. 19). By Domesday times, wood had disappeared neither from these clay areas of the upland nor from those of the fen islands (see Fig. 17). Earlier evidence is not wanting. The most common woodland terms found in English place-names are *-ley* "a wood" or, later, "a clearing"; *hay* "enclosure", and often "enclosed wood"; *stubbing* "clearing"; and *stocking* "land cleared of stocks". The clay areas of southern Cambridgeshire each contain parishes with names ending in *-ley*.[1] In these, as well as in neighbouring parishes, other names have survived, as well as a number of field-names containing all four of the above elements. In the intervening belt of chalk, there is a solitary lost *ley* in Little Wilbraham, another in Fulbourn, and "The Leys" in Burwell—a marked contrast with Ashley-cum-Silverley, on the eastern clay, with its four additional examples of *ley*, two of *hay*, and one of *stubbing*.

Croydon Wilds and Hatley Wilds, on the western clayland, are names of particular interest in connection with Cambridgeshire woodland. The first occurs in 1285 as *in Waldis de Craudenn'* and, as late as 1760, as *Croydon Wold*; the second is found in 1277 as *in Weldis subtus boscum de Hayley*, i.e. Hayley Wood in Little Gransden. Both are on high ground and both names contain *weald, wald*, used in Old English of forest-land, especially of high forest-land. Other evidence of the wooded nature of the ground is to be found in the names Hatley and Hayley, and in *Longehay* and *Dreyhirst*, in Little Gransden. Farther north, there was a *Woldeslande* and a *Grenewold* in Elsworth, while Dry Drayton was formerly called *Walddraiton*. Between Croydon and Elsworth there are references to *Berstunesweald* and *Kakestunesweald* in Caxton and *in Waldis de Brune* (i.e. Bourn). There can thus be little doubt that this district was once known simply as *Weald* or *Wold*.

The clay islands of the fen were also well wooded. The first element in Chatteris and Chettisham is probably the British *cet* "wood". Near Chatteris is Langwood Fen, whilst medieval woodland is frequently mentioned at Chettisham. On these clay islands there are examples not only of the terms already discussed, but also of *hyrst* and *holt*, two other names for a wood. The last two elements are often difficult to recognise without early forms. *Holt* occurs in Singlesole at Thorney; Throckenholt

[1] In the east are Ashley-cum-Silverley, Brinkley, Cheveley and Westley Waterless. In the west are Childerley, Eltisley, Graveley, East Hatley, Hatley St George and Madingley.

in Parson Drove; and Apes Hall in Littleport; *hyrst* in Shrewsness Green in Upwell; and Boleness in Wisbech St Peter. The occurrence of *hyrst*, *leah*, and *holt* on the peat and silt is noteworthy.

From the Isle of Ely, too, comes an example of *wold* covering a wide area in which other names indicative of the former existence of woodland are to be found. In early medieval documents, numerous references are found to a place *Walde* or *Wolde* in Witchford. According to Bentham, the name *Wold* survived in his day as that of certain arable and pasture lands in Witchford,[1] but it was undoubtedly once used of an extensive district. In the west of the Isle, at Sutton, was a *hythe* or landing-place in the wood known as *Waldhethe*, and in the same parish there were two hills called *Waldun* and *Waldelowe*, while a Woolden Lane still survives in Haddenham. From Witchford, this forest-land stretched into Ely where both *le Wold* and *Woldeffeld* are mentioned, whilst in Downham, too, there is *Brodwold*. A road or track called *Waldehethewey* ran across this *wold* which must have included most of the high clayland of the Isle of Ely; part of this, at least, in Wilburton and Witcham, was known as *Bruneswold* at the end of the thirteenth century.

Much more that is of topographical and historical interest may be gathered from the place-names of the Isle of Ely. Seadike Bank and Sea Field in Leverington recall the memory of the sea-wall that once protected the coast of Cambridgeshire from Tydd St Giles to Wisbech, and that gave name in Norfolk to Walsoken, West Walton, and Walpole. Coveney, "the island in the bay", is a reminder that the West Fen was once marsh and water; the coast of the ancient bay can easily be traced from the contours. Not far away is Wardy Hill, "the island from which watch was kept", a name that may be of some historical significance for it is on the line of the long, narrow island running north to March, "the boundary", probably that between East and Middle Anglia. Among the other numerous islands are Shippea and Quy where sheep and cows were pastured, Henny and Cranney, frequented by wild fowl and herons, and Manea near where the parishes of the neighbourhood pastured their cattle in common.[2]

Around the island ran innumerable watercourses, the Old and the South Eau, unetymological Gallicisings of the Old English *ēa* "river", already noted in the Rhee (or Cam) and surviving also in Welney and Wissey. Bradney House in Benwick, and Bradney Farm in March, are both near

[1] J. Bentham, *The History and Antiquities of the Conventual and Cathedral Church of Ely* (2nd ed. 1812), p. 75.

[2] Manea means "island or low-lying land held in common or where commoning took place".

the old course of the Nene, here once known as *brādan ēa* "the wide river". The extensive area known as Byall Fen was formerly called *Byee* from a river which formerly flowed across the fen from Chatteris to Downham.[1] Very common, too, are the terms *lode* and *gote* (meaning some kind of stream), *ditch*, *delph*, used of an artificial watercourse, and *lake*, "a sluggish stream".

Of the numerous fenland *meres*, Whittlesey Mere and Soham Mere were probably the largest. But the names of others survive, and many more are lost. With the draining of the fens, these meres became marshland and now often appear as *moors*. Redmoor was "the reed-mere", Gosmoor was frequented by geese, and Foulmire was the home of wild fowl. Fisheries, too, were common, and were called *weirs* as at Upware. There are many references also to landing-places, or *hythes*, not always easy to recognise in their modern form, e.g. Horseway, Willey, Aldreth, Swavesey, and Witcham Hive. The old industries of the fens, digging for peat (always called turf by the fen folk), the cutting of sedge for thatch, and the growth of fodder for cattle, are commemorated in the common names Turf Fen, Sedge Fen, Fodder Fen, and Mow Fen. The Joist Fen at Waterbeach was one in which cattle were *agisted*, i.e. allowed to feed for a fixed rate per head.

Many fenland names owe their origin to the exigencies of draining or commemorate the reclaimers of the fens. Adventurers' Land and The Undertakers represent part of the land assigned to the Earl of Bedford (after whom the Bedford Level is named) and his associates in return for their *undertaking*, and *adventuring* upon, their immense task.[2] "The Lots" preserve a common term "the lot or dole" used of the allotment of land in the fens, while Lockspit Hall owes its name to the lockspits or small trenches used to divide these lots. Cradge Bank was so called because it was backed with clay to prevent water from trickling through. Stampfen Drove, Gravel Dike, and The Stacks, preserve local names for "letts or impediments hindring the fall of the waters". Other terms of interest are found, particularly in names now lost, but with these there is not space to deal. Sufficient evidence has, however, been given in this brief and incomplete survey to show that the place-names of Cambridgeshire and the Isle of Ely are full of interest and that the nomenclature of each locality has its peculiar characteristics.

[1] This is marked on Saxton's Map of Cambridgeshire (in W. Camden's *Britannia*, 1607) as *The fyrth dyck*, so called, no doubt, from Doddington *Frith*.

[2] For "Undertaker" and "Adventurer" see p. 181 below.

CHAPTER EIGHT

# THE VILLAGES OF CAMBRIDGESHIRE

By John Jones

THE VILLAGES OF CAMBRIDGESHIRE NUMBER ABOUT ONE hundred and sixty.[1] There are, in addition, some half-dozen urban areas, but many of these only contain overgrown villages. Nearly all of them are mentioned in the Domesday Book. It is certainly true to say that the village geography of the County has been stable through the centuries. Fig. 22, showing the distribution of Domesday settlements, represents quite well the pattern, if not all the detail, of the present-day village distribution as shown in Fig. 23. The main difference between the two maps is the addition of a number of villages upon the silt area of the northern Fenland.[2]

## LOCATION OF VILLAGE SITES

Even a cursory glance at Fig. 23 shows that these villages are not evenly distributed. Three stretches of country appear villageless: a district in the north of the County; an east-west strip south of the Ely cluster of villages; and a north-east—south-west strip in the south-eastern quadrant of the County. On the other hand, the area with the most dense distribution of villages occupies the south-western quadrant.

Four physical features are reflected in this variation: the Fenland and its islands; the upland with its contrast between clay and chalk; the river valleys; and the narrow strip of country between fen and upland which may be conveniently designated the "fen-line". From the point of view of location, therefore the villages fall into four groups:

(1) Fen villages.
(2) Upland villages.
(3) Valley villages.
(4) Fen-line villages.

Naturally, these groups are not mutually exclusive, for some sites claim admission to more than one group. Take Cambridge town itself, for example. It is on the fen-line, but it is also in the valley of the Cam, and

[1] For a fuller account, see J. Jones, *A Human Geography of Cambridgeshire* (1924).
[2] For a discussion of these differences, see H. C. Darby, "The Domesday Geography of Cambridgeshire", *Proc. Camb. Antiq. Soc.* xxxvi, 35 (1936).

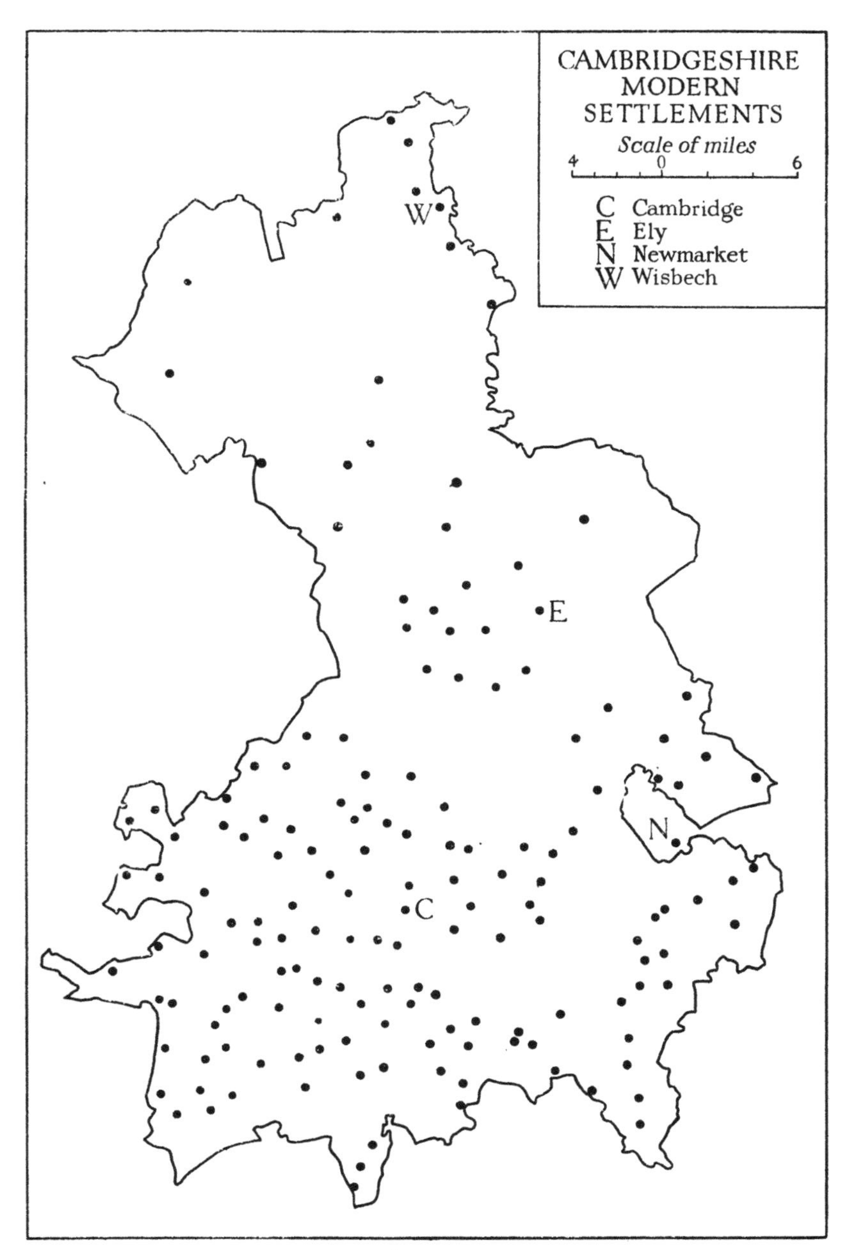
CAMBRIDGESHIRE
MODERN
SETTLEMENTS
Scale of miles
4
0
6
C Cambridge
E Ely
N Newmarket
W Wisbech
W
E
N
C

Fig. 23.

owing to the latter fact it has outstripped in size the other fen-line settlements.

*South-eastern Cambridgeshire.* Fig. 24 gives details of the country between Cambridge town and the eastern County boundary. It covers an area of 170 square miles, and all the village sites and watercourses are inserted.

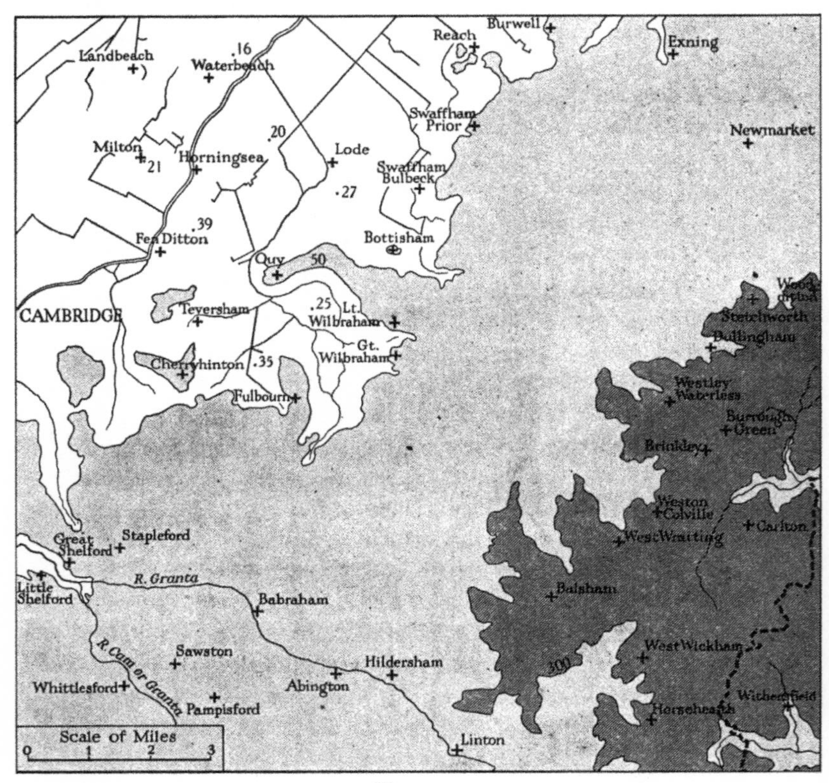


Fig. 24.

Villages to the east of Cambridge.

The figures in the area below the 50 ft. contour are spot heights above O.D.

The relief is indicated by the 50 ft. and 300 ft. contours. At the time of the settlement, this part of the County most probably included three zones of vegetation. To the north-west, there was a marsh of reeds and rushes, interspersed with patches of better drained land. The country between the two contours was a dry area, with large open spaces of grass, dotted here and there with hawthorn bush, and probably with some beech woods. The 300 ft. contour line agrees very closely with the edge of the Boulder

Clay which overlies the Chalk on the top of this upland area (see Fig. 29), and which, most probably, was covered with deciduous woodland (see Fig. 17).

The outstanding feature of this district to-day is the considerable area, running north-east—south-west, that is villageless and also streamless until the River Bourn is reached in the south. This belt is the wide strike exposure of the porous chalk country. In its completely dry character lies the explanation of its unpeopled state. To the south, near the Bourn and Granta rivers, come the valley villages. Some of them have "ford" terminations; in each case, the precise site was determined by rising ground safe from flood. In the east, very near to the 300 ft. contour and between woodland and open country, are the upland villages. Until recently, some of these villages have used for domestic purposes the surface rain-water that drains into hollows on the impervious clay; wells have to be sunk so deep before reaching the bottom of the chalk, that they are expensive items. Along the edge of the clayland, then, most of the upland villages are situated. The advantages of a site on the edge of the wood, rather than within it, can be readily appreciated. Finally, the 50 ft. contour marks, roughly, the junction between chalk and fen. Along this line, springs gush out from beneath the chalk, and so, following the fenland edge, is a string of eleven villages from Burwell in the north to Cherry-hinton in the west. To the north-north-west of these fen-line sites, recti-linear watercourses indicate the drained fen. There, on the drier spots, are other marginal villages; a few spot heights have been inserted.

*South-western Cambridgeshire.* Fig. 25 gives details of villages and streams in the country to the west of Cambridge town, and it covers the same acre-age as Fig. 24. The Cam Valley occupies the south and east of the map. In the middle west is a small plateau over 100 ft. above sea-level; this contains two ridges (above 200 ft.) that run east-west to form the boundaries of the Bourn Brook Valley. The surface of the plateau is almost entirely composed of clay of various kinds, mainly Boulder Clay and Gault, but also Kimeridge and Oxford Clays (see Fig. 29); presumably it was wooded in early times.[1] Generally speaking, the plateau forms the watershed between the tribu-taries of the Ouse and those of the Cam; many of the streams marked on the map are only a foot or two in breadth, but, even so, their presence shows this to be a district quite different from that on the eastern side of the County.

The top of this plateau is almost villageless. The villages are set around the edges. The elements important in their distribution seem to be as follows:

(1) The valley sites of the Bourn Brook and of the Cam explain

[1] For Domesday Woodland, see Fig. 17 above; for earlier evidence, see p. 103.

themselves; the Hatley's and the Gransden's to the west belong also to this category.

(2) With a west-to-east dip, the lower limit of the Chalk (i.e. the spring-line) is at a higher level than in the east of the County. To some extent, it coincides with the 100 ft. contour, but it does rise to 200 ft.

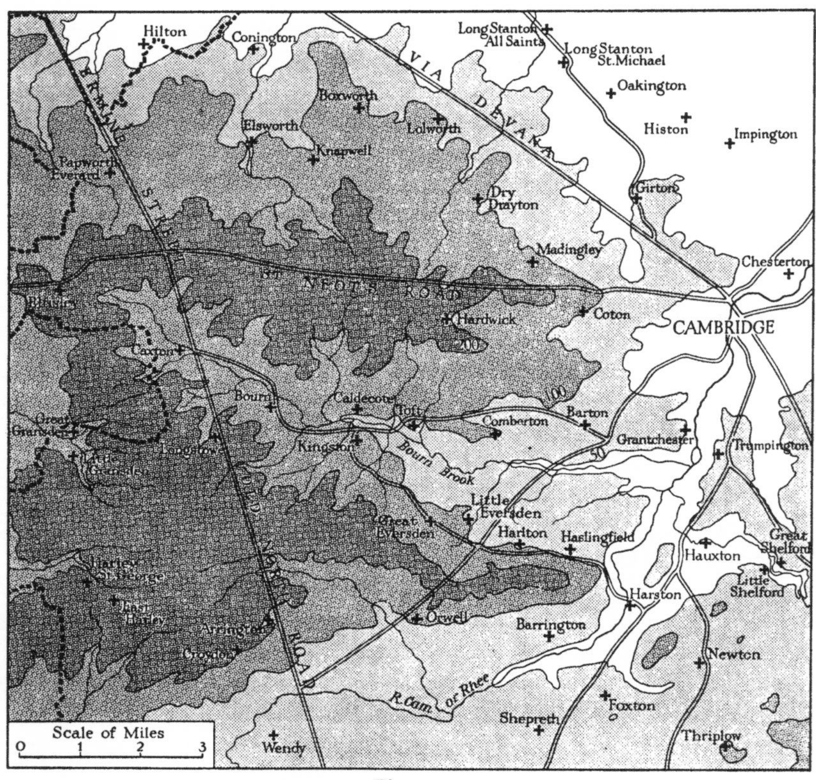


Fig. 25.

Villages to the west of Cambridge.

Following interrupted outcrops of the Chalk, along the eastern and southern edges of the plateau, are the villages of Madingley, Coton, Barton, the Eversden's, Harlton, Haslingfield, Barrington, Orwell, Arrington, and Croydon.

(3) In the north, the village sites still follow the 100 ft. contour, which roughly marks the junction between the Boulder Clay and a varied series of Gault, Lower Greensand, Kimeridge Clay, and Ampthill Clay. It is interesting to note that these upland villages lie off the *Via Devana*.

(4) To the north, and outside the map, come the fen-line villages—Cottenham, Rampton, Willingham, Over, Swavesey. These lie much lower than the corresponding villages in the east of the County, where the chalk escarpment borders the peat and brings the 50-ft. contour near to it.

*Northern Cambridgeshire.* In the undrained Fenland the islands were the critical sites determining settlements. An eighth-century monk, Felix, records that:

> There is in Britain a fen of immense size, which begins from the river Granta [*Grante*] not far from the city, which is named Grantchester [*Granteceaster*]. There are immense marshes, now a black pool of water, now foul running streams, and also many islands, and reeds, and hillocks and thickets, and with manifold windings wide and long it continues to the north sea....[1]

By the eleventh century, however, as the Domesday map shows (Fig. 22), the Fenland was not without villages. But settlement was prohibited upon the peatlands because the soil provided no stable foundations on which to build. Judged by the analogy of later times, even those portions that escaped winter flooding were subject to an annual heaving motion as the swollen peat absorbed more and more water. Consequently, not one Domesday village was located in all the peat area, with the sole exception of Benwick, and there only because a local gravel substratum approached within a few inches of the surface.[2] The open unoccupied area to the north of the County, shown on Fig. 23, is an expanse of peat. So is the east-west strip of country south of the Ely cluster of villages; here, in pre-drainage days, the Old West River carried part of the Ouse around the island of Ely and hence to the Wash.[3] The villages were all upon the islands. But the silt area, to the north, was composed of a substance more solid than fen peat, and offered better opportunities for continuous settlement. It is true that, in Domesday times, Wisbech alone stood here, but on the modern map, the silt area bears a number of additional villages.

## PARISH BOUNDARIES

Not only the sites of villages, but also the size, shape, and boundaries of parishes, are related to the geographical circumstances of a country. Fig. 26 shows the modern parish boundaries in Cambridgeshire. The parishes vary considerably in size. There are eleven parishes each containing less than

[1] Felix, *Life of St Guthlac* (Anglo-Saxon version), edited by C. W. Goodwin (1848), p. 21. Grantchester here refers to Cambridge itself.
[2] S. B. J. Skertchly, *The Geology of the Fenland* (1877), p. 4.
[3] See footnote 1, p. 183 below.

1000 acres—all in the southern part of the County; while ten parishes cover more than 10,000 acres each—all on the Fenland. This latter fact arises from the scarcity of sites in the fens. Thus, just below the northernmost part of the County, Whittlesea and March together stretch 15 miles across a part of the County whose total width is only 17 miles. Whittlesea and March stand 26 ft. and 20 ft. respectively above sea-level, and in this district there is now no other spot more than 10 ft. above sea-level.

The parish boundaries are partly natural and partly artificial. The Cam itself is a good example of the former. It is used as a boundary for almost its entire length from its source to where it joins the Old West River (i.e. the Ouse) in the Fenland. Its tributary, the Bourn Brook, coming from the south-western plateau, also forms the boundary between many parishes, despite the fact that it is of no great width; on the other hand, the wider River Bourn, coming from the south-east, flows right through the middle of several parishes. In the Fenland, many parish boundaries are so straight because they follow artificial watercourses; and in some cases, apparently, adjustments were made when the drains were cut.

In a county of low relief, there cannot be many boundaries fixed by a crest-line, but there is one well marked in the south-west, and lettered A–B on Fig. 26. It runs along the ridge separating Bourn Brook from the upper Cam Valley. This is also the line of an old trackway, the Mare Way, leading from Ermine Street to the Cam Valley at Harston. The line C–D is not exactly along the crest of the ridge to the north of Bourn Brook, but it is very nearly so. In any case, it runs along the Cambridge-St Neots Road, and although this is not usually claimed to be an ancient way, still it does seem to date from the time when boundaries were being established. The *Via Devana* also formed parish boundaries during a great portion of its length (line E–F). It is part of the Roman road from Colchester to the Midlands. From where it enters the County at E, for a distance of 11 miles, it separates parishes; and it continues this function to the north-west of Cambridge along the line G–H, which here forms the main road from Cambridge to Huntingdon. Ermine Street, also, helped to define boundaries (line K–L). Probably the oldest track in the County is the Icknield Way,[1] which is represented to-day by a section of the London-Newmarket road (line M–N). The parish boundaries, it is true, do not strictly follow the road, but that does not indicate that the way was not in existence when the boundaries were established.[2] Another straight boundary that attracts attention on Fig. 26 is the line P–Q. This is not a road, but the Devil's Dyke, that is a parish boundary for 10 miles.

[1] See p. 84 above.
[2] See Fig. 19.

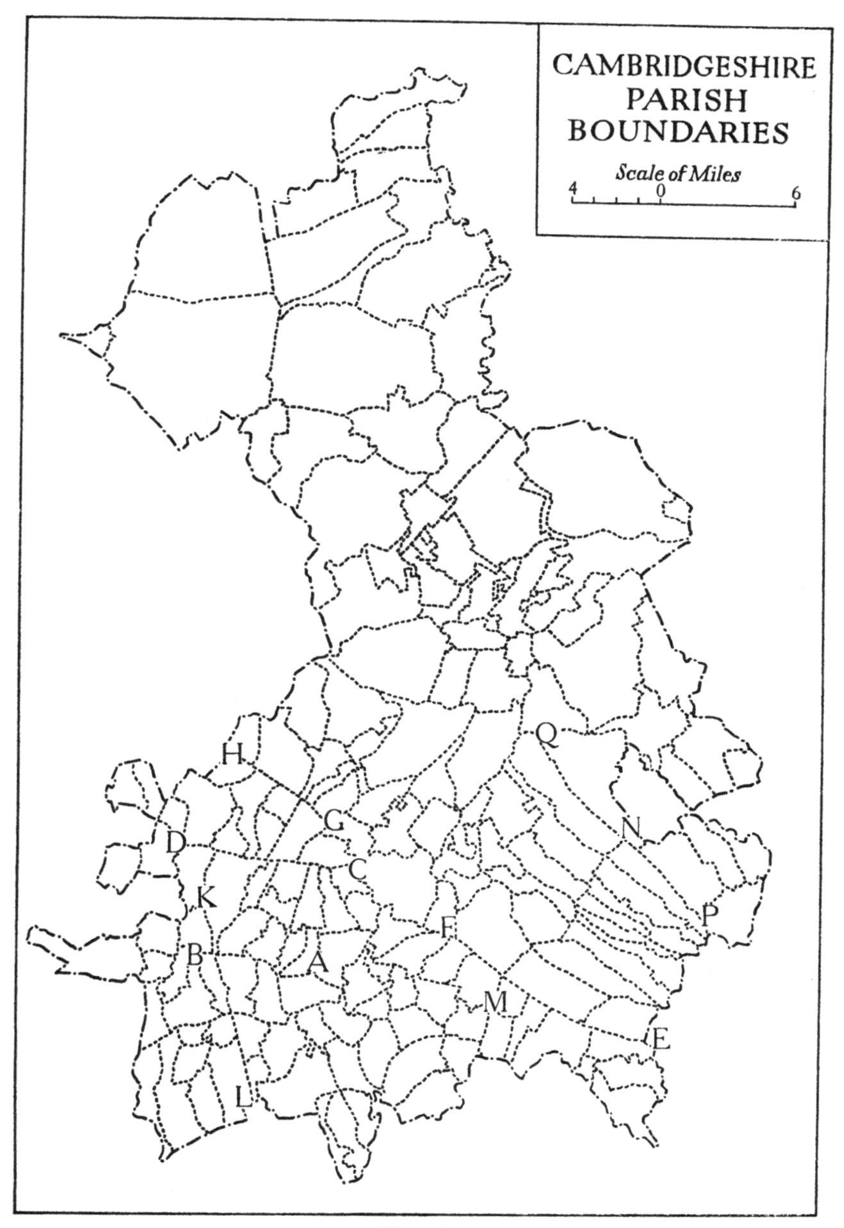


Fig. 26.

The letters A–Q refer to the sites named on p. 112.

The shapes of parishes are so varied that they defy classification. The chances of topography and of time cannot be reduced to generalisation. But the configuration of parishes in the south-east of the County certainly stands out as something peculiar. The upland villages of Fig. 24 are seen, on Fig. 26, to form a group of long narrow parishes lying side by side. They are oriented down the slope and not along it. This arrangement gives to each parish a variety in soil and vegetation. If the parishes had been, say, square, any single one might have contained nothing but clay (and woodland), while its neighbour could have consisted entirely of open chalk down. Beyond the boundary M–N, there is a second tier of parishes, lower down the slope. This side-by-side arrangement secures for them, also, two types of terrain—an area of chalk and an area of fen. South of the line E–F the parishes are oriented at right angles to the *Via Devana*; this gives to each a stretch of chalk upland as well as a share in the valley alluvium. A somewhat similar arrangement is found in the parishes of the Bourn Brook Valley.

All these indications of order in the parish map are but stray hints and glimpses; from them, however, we can see reason at work when the early settlers laid down the foundations of the present village geography of the County.

## PARISH CHURCHES

The prominent centre in every village was the parish church, and it is interesting to see how the churches, like their parishes, reflect the geographical circumstances of their environment. The four maps of Fig. 27 show the main types of material used in their construction; naturally this classification is based only upon the chief materials used. Very often, a clunch or flint church has imported stone pillars and arches. For it should be remembered that Cambridgeshire is not very rich in good building stone. Even so, facilities of transport have not destroyed the local background. Flint churches are located mainly along the southern boundary of the County, where the flint was obtained from the Upper Chalk (see Fig. 4). Clunch, out of the Lower Chalk, is easily weathered, but, in the absence of better stone, it has been used to help build a number of churches. Rubble, probably consisting of stones gathered from the Boulder Clay and mixed with mud, has been used in thirty churches, but these often have stone dressings. The remaining churches are built of imported stone; most of the fen churches come within this category, for transport was comparatively easy along the fen waterways.

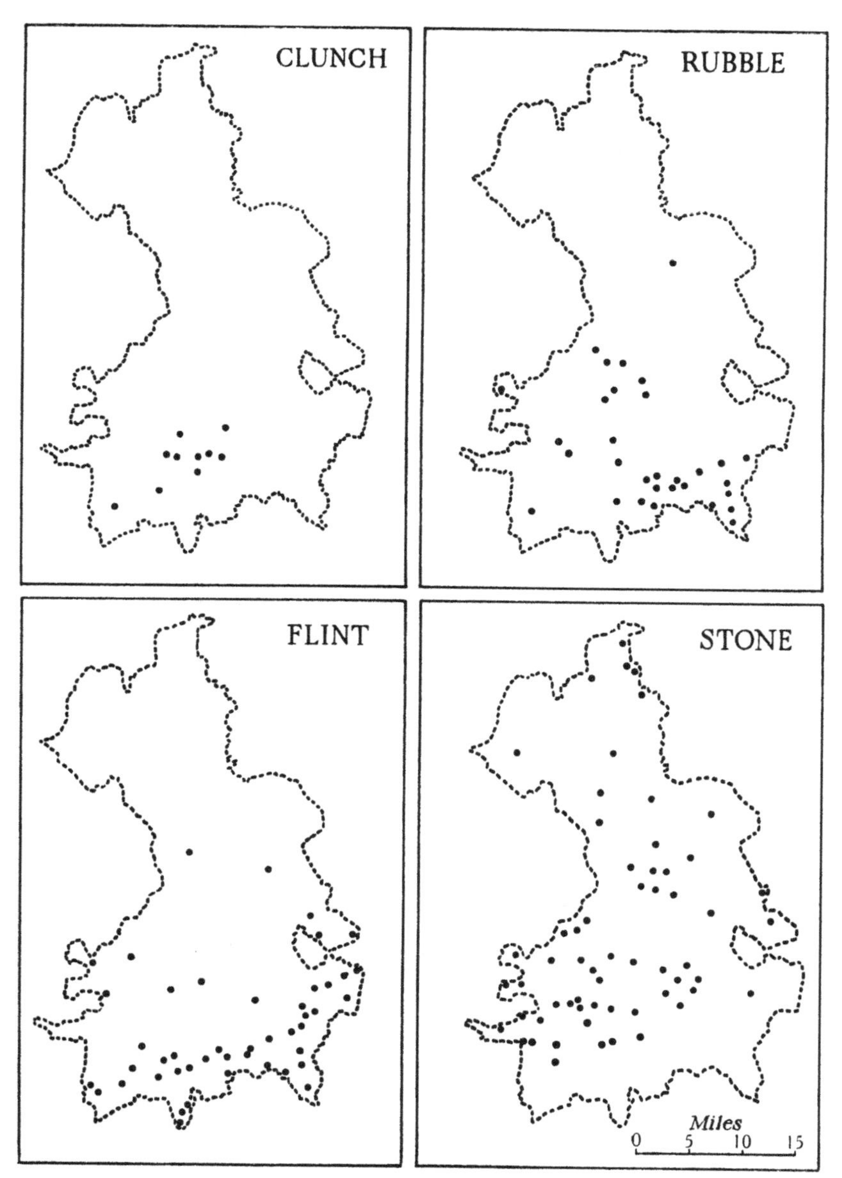


Fig. 27.

Cambridgeshire: Building materials used in churches.

CHAPTER NINE

# CAMBRIDGESHIRE IN THE NINETEENTH CENTURY

By H. C. Darby, M.A., PH.D.

A PICTURE OF CAMBRIDGESHIRE ABOUT THE YEAR 1800 CAN be gained from two reports made to the Board of Agriculture. Both have the same title—the *General View of the Agriculture of the County of Cambridge*. The first was written by Charles Vancouver, and was published in 1794. The second was written by W. Gooch, and, although its Preface is dated 1807, the book itself was not published until 1813. Taken together, these two surveys supplement one another to provide an outline of the main features of the geography of the County in 1800.

Both reports are accompanied by what is substantially the same map of land utilisation (Fig. 28). Comparison with the geological map of the County (Fig. 29) provides an explanation. The outstanding contrast was between the northern Fenland and the southern upland. Conditions in the upland area reflected directly the geological division. Between the "close heavy, compact Clay" in east and west, the belt of chalk country stood out, running south-west—north-east. Where the chalk outcrop was almost waterless, there was "heath"; while the "valley through which the river Cam flows to Walton, is chiefly laid out into dairy farms, and hence it has its name, i.e. the Dairies".[1] In the Fenland, to the north, the islands stood out. Of the peat fens around, some were "under cultivation", others were "drowned or waste". Finally, the silt area of the extreme north maintained its reputation as "rich pastures".

One of the main objects in the making of these reports was an enquiry into measures necessary for improvement. In Cambridgeshire, there were two agricultural controversies that reflected the geographical circumstances of the time. One was associated with the need for an improved drainage; the other with the need for an increased enclosure of the common open-fields.

Vancouver estimated[2] the total acreage of the County as 443,300, divided as shown in the following table. The correct area of Cambridgeshire is 553,555 acres; but, even so, the proportion between the

[1] C. Vancouver, p. 87. Walton, on the north bank of the river, and lying to the south-east of Orwell, is marked on John Cary's map of 1818.

[2] *Ibid.* p. 193.

different districts, and their relative values as given by Vancouver, are probably accurate enough to be indicative.

| Description of land | Number of acres | Rental or value per acre |
|---|---|---|
| | | £ s. d. |
| Enclosed Arable | 15,000 | 18 0 |
| Open Field Arable | 132,000 | 10 0 |
| Improved Pasture | 52,000 | 1 0 4 |
| Inferior Pasture | 19,800 | 10 9 |
| Wood Land | 1,000 | 15 0 |
| Improved Fen | 50,000 | 15 0 |
| Waste and Unimproved Fen | 150,000 | 4 0 |
| Half Yearly Meadow Land | 2,000 | 12 6 |
| Highland Common | 7,500 | 10 0 |
| Fen or Moor Common | 8,000 | 3 0 |
| Heath Sheep Walk | 6,000 | 2 6 |

## DRAINING THE FENLAND

The condition of the Fenland towards the end of the eighteenth century was far from satisfactory,[1] and the phrases that Vancouver used to describe the fen parishes were monotonously similar. The fens of Fordham were "in a very bad state"; those of Bottisham were "in a deplorable situation, and subject to frequent inundations"; those of Burwell, too, were "constantly inundated". So were those of Ely and Upwell and Outwell. At Elm, cultivation was "very uncertain"; at Littleport, it was "extremely precarious". And so the tale of woe continued throughout the whole of the Isle of Ely. Only in a few parishes were the fens "tolerably well drained"; and even then often at "very considerable expense". The tragedy was all the greater because some parts of the fen under improved cultivation yielded "a produce far beyond the richest high lands in the county".[2]

These deplorable conditions were attributed by the men of the time to the "want of a better outfall through the haven of Lynn".[3] Within 3 miles of Lynn, the Ouse made a great bend following a course of about 6 miles or so. The channel of this river was of varying width and was full of shifting sandbanks. In some places it was as much as a mile wide, comprising a number of uncertain streams; and, during floods, the flow of the river was much impeded.[4] Many people believed that the only solution

[1] See p. 187 below.

[2] C. Vancouver, pp. 202–3. This estimate was based upon conditions in 50,000 acres of fen around "Chatteris, Elm, Leverington Parson Drove, Wisbich, St Mary's, and Thorney". Compare with the general information given in Vancouver's table above.

[3] *Ibid.* p. 139.

[4] See p. 191 below.

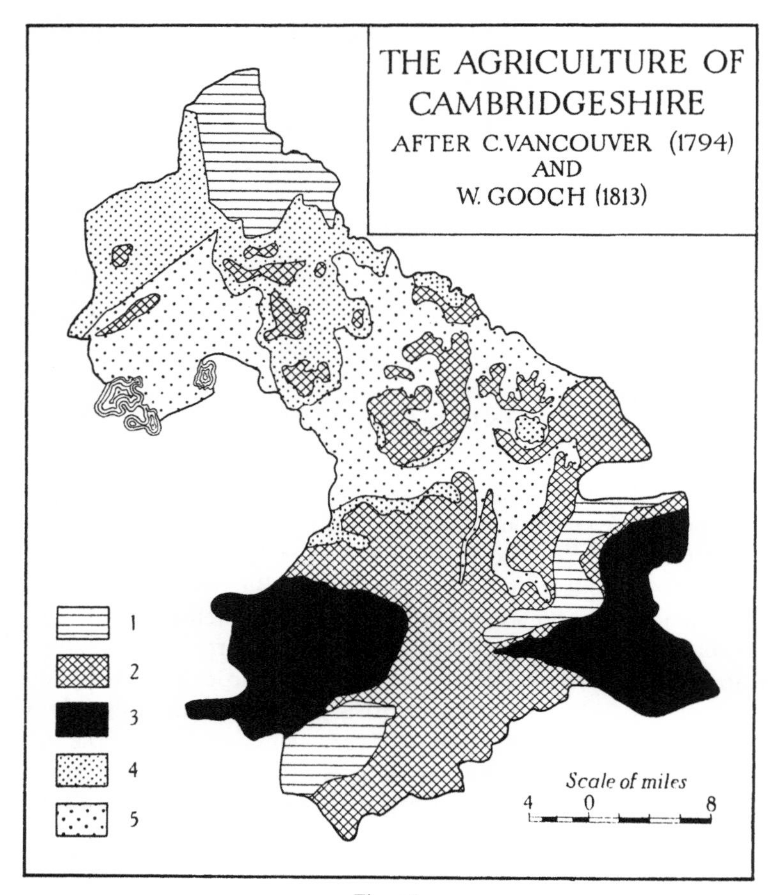


Fig. 28.

Land Utilisation in Cambridgeshire about A.D. 1800.

1. Part of Newmarket Heath, the Valley called the Dairies; and rich pastures produced from the Sea around Wisbeach.
2. Chalky, Gravelly, Loam and tender Clay.
3. Close heavy, compact Clay upon a Gault.
4. Fen under Cultivation and in the High Land Sand.
5. Drowned or waste Fen but all very capable of being reclaimed.

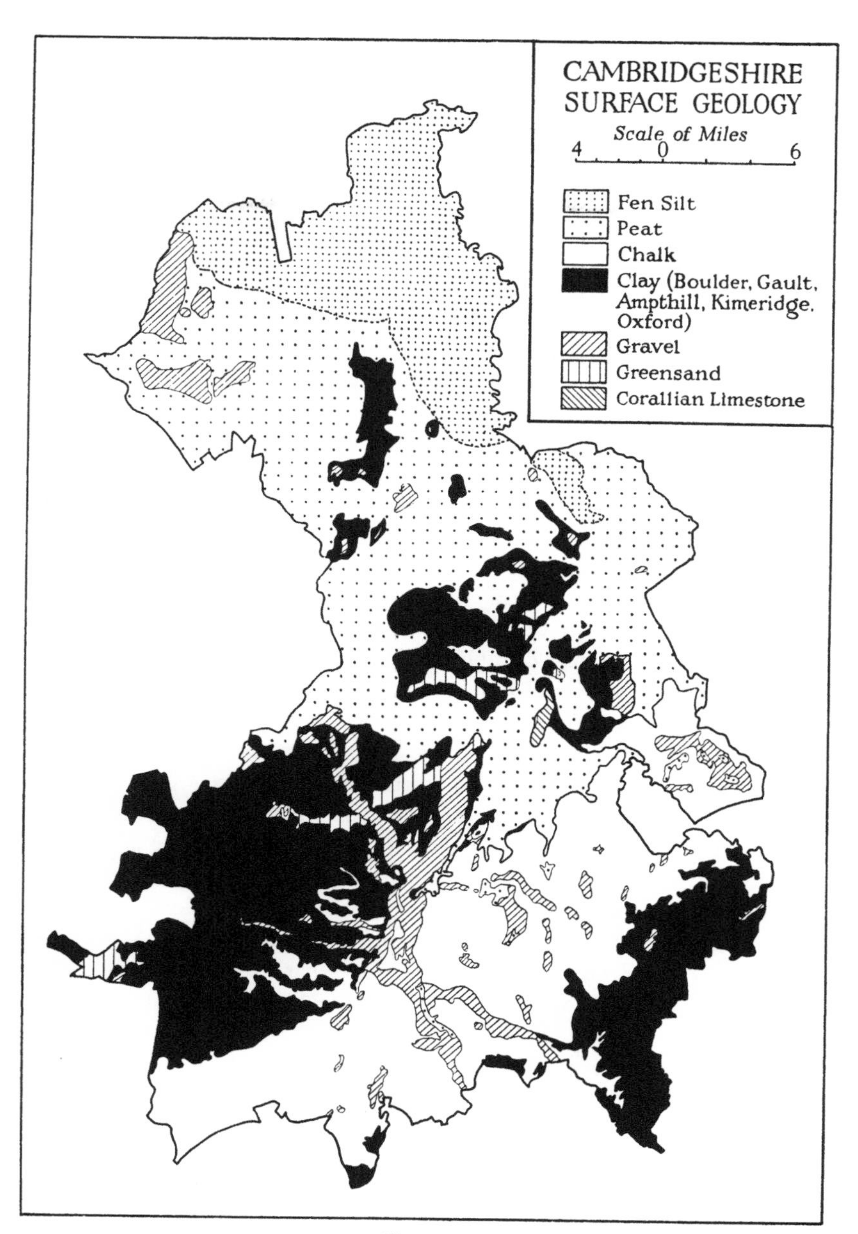


Fig. 29.

This map is based upon that of the Geological Survey. The boundary between the fen silt and peat is taken from the map accompanying S. B. J. Skertchly's *The Geology of the Fenland* (1877). This boundary is "very obscure, for the peat thins out insensibly..." (p. 129).

was to eliminate the great bend by a new straight cut made large enough to take quickly out to sea the whole body of the Ouse water. The Eau Brink Act, authorising the cut, was obtained in 1795. Disputes and delays followed, and the work was not completed until 1821.

It was therefore in an atmosphere of controversy that the two reports of 1794 and 1813 were compiled. Indeed, one of Vancouver's main objects was to see "how far the proposed measure, of diverting the course of the river Ouze, from its present channel between Eau-brink, and the Haven of Lynn, would embrace all the objects so fondly anticipated by the promoter of that measure".[1] Gooch, too, dealt with the desirability of an Eau Brink Cut. As he pointed out, the opinions were varied enough. But, at any rate, "all were agreed that something must be done or the country will be lost". During the succeeding years of the century, the straight line of the Eau Brink Cut was continued seawards by the Marsh Cut of 1852, and, later, by training walls built to carry the Ouse waters out to sea amid the shifting banks of the estuary.[2]

Complementary to the outfall question was the problem of the internal drains. Internal drainage during the eighteenth century had been accomplished by windmills that pumped water from the lowering peat surface into the high-riding river channels.[3] But, in the report of 1813, there is a hint of changes to come. The mills, depending on wind, were "often useless when most wanted",[4] and the proprietors consequently sustained "material injury". To remedy this, steam engines had been recommended, and, declared Gooch, "I found many persons in the county entertaining an opinion that they would answer". It was argued that the advantages to be obtained from the introduction of steam to the fen country were "almost incalculable". But there was delay and hesitation. Not until 1820 was the first steam-driven mill set up at Bottisham Fen. Succeeding years[5] but verified the prophecy of Gooch that "until a power can be commanded at will, for the drainage of the fen-country, it can never attain its full prosperity".

The improvements in draining that marked the nineteenth century were paralleled by improvements in agricultural practice. Owing to continued shrinkage and wastage, the surface layer of peat in the southern Fenland was becoming so thin, that, in some districts, the underlying clay was within easy reach of the plough.[6] Thus the virtues of clay were discovered,[7] and so potent did they prove to be that, where the clay was too

[1] C. Vancouver, p. 8.
[2] See pp. 191 and 201 below.
[3] See p. 187 below.
[4] See W. Gooch, pp. 239 *et seq.* for the quotations that follow in this paragraph.
[5] See pp. 188–9 below.
[6] See pp. 131 and 186 below.
[7] J. M. Heathcote, *Reminiscences of Fen and Mere* (1876), p. 90.

deep down for the plough, the practice of digging for it became frequent. In 1811, there was being advocated the use of the "most excellent clay marl" that underlay "the greatest part if not the whole of the Bedford Level".[1] By 1830, the practice was "so very modern" that Samuel Wells found "some difficulty in giving an accurate account of its singular process".[2] But the practice had come to stay; so much so that, in 1852, J. A. Clarke could write:

> Within the last 30 years the system of digging and throwing up this clay where it is too deep for the plough has been introduced into universal operation. The new husbandry quickly extended itself: farmers may be cautious of new improvements, but this was too obvious for dispute, too near at hand for refusal.[3]

Wherever clay could be found tolerably near the surface, "claying the land became the acknowledged mode of cultivation in the Fenland".[4] The peat lands thus became "the most productive of soils yielding the most luxuriant crops of wheat, oats, coleseed and turnips".[5]

## ENCLOSING THE COMMON-FIELDS

Of the 147,000 acres of arable land in Cambridgeshire in 1794, 132,000 acres lay in open-fields, and followed the traditional open-field husbandry of the English plain. Of course it is possible that some of the 52,000 acres of "Improved Pasture", recorded in Vancouver's estimate, included land laid down to grass on enclosure. But even with this allowance, there can be no doubt about the unenclosed character of the Cambridge countryside. Corroboration is provided by the fact that of the ninety-eight parishes described in detail by Vancouver, eighty-three were still open; only fifteen had been enclosed. And Vancouver considered that no improvement was possible until the intermixed strips "dispersed in the common open fields" had been brought together into compact holdings. Enclosure appeared "to be indispensably necessary" and urgent.

> "I have made it my particular care", he wrote, "to mix and converse with the yeomanry of the county, and in their sedate and sober moments, to possess myself fully of their experience, and local knowledge; and finally to ascertain the general sentiment as to this important innovation upon the establishment of ages."[6]

In some places, people were doubtful; thus at Teversham the idea of enclosing was "not all relished".[7] In other places, "the most thinking

[1] R. Parkinson, *The Agriculture of the County of Huntingdon* (1811), p. 299.
[2] S. Wells, *History of the...Bedford Level*, i, 442 (1830).
[3] J. A. Clarke, *Fen Sketches* (1852), pp. 244–5.
[4] J. M. Heathcote, *op. cit.* p. 90. See p. 152 below.
[5] J. A. Clarke, "On the Great Level of the Fens", *Jour. Roy. Agric. Soc.* viii, 92 (1848).
[6] C. Vancouver, p. 195.
[7] *Ibid.* p. 47.

farmers"[1] were very much in favour of "the laying of the intermixed property together in the open fields".[2] And Vancouver was emphatic in demonstrating the improvement in crop yields that resulted from enclosure.

Want of enclosure was also felt in the Highland Common which "in severalty" would have been doubled in value; while the Half Yearly Meadow Land, "dispersed through the hollows of the open fields", would even more than double in value "by proper draining and being put into severalty".[3]

The report of 1813 by W. Gooch showed that "most of the arable husbandry of this county" was still foreign "to present practice in the best cultivated countries".[4] Many people still believed that the older methods were the best, and "this bigotry" was widely spread. But something had certainly been done to redeem the County "from the imputation it has so long lain under, of being the worst cultivated in England".[5] By 1807, the open-field arable was "much lessened", and a great part of "the waste and unimproved fen, half-yearly meadow, highland common, fen or moor common, sheep-walk heath", had become enclosed arable and pasture. In the case of open-field conversion, the total rental had more than doubled: on other lands it had trebled at least.[6]

But more still remained to be done. In 1822, when William Cobbett travelled along the Old North Road from Royston to Huntingdon, much of the country was still treeless and hedgeless, full of "those very ugly things, common-fields", and looking "bleak and comfortless" to the eye.[7] Still later, in 1830, between Cambridge and St Ives, Cobbett again saw "open unfenced fields".[8] But Cambridgeshire was coming into line with the rest of the English plain. By 1847, all its open common-fields, "with the exception of five or six parishes",[9] had been enclosed.

## THE CURVE OF PROSPERITY

The fluctuations of agricultural fortune in Cambridgeshire during the nineteenth century reflected, very largely, variations in the prosperity of

[1] C. Vancouver, p. 53. [2] *Ibid.* p. 147. [3] *Ibid.* p. 204.
[4] W. Gooch, p. viii. [5] *Ibid.* p. 56. [6] *Ibid.* p. 2.

[7] W. Cobbett, *Rural Rides* (Everyman's edition), i, 80–2: "Immediately upon quitting Royston, you come along, for a considerable distance, with enclosed fields on the left and open common-fields on the right....The fields on the left seem to have been enclosed by act of parliament; and they certainly are the most beautiful tract of *fields* that I ever saw. Their extent may be from ten to thirty acres each. Divided by quick-set hedges, exceedingly well planted and raised" (p. 80).

[8] *Ibid.* ii, 236.

[9] S. Jonas, "On the Farming of Cambridgeshire", *Jour. Roy. Agric. Soc.* (1847), p. 38. G. Slater gives nine parishes enclosed after 1847 (*The English Peasantry and the Enclosure of the Common Fields* (1907), p. 273). But for at least ten, see E. M. Hampson, "Cambridge County Records", *Proc. Camb. Antiq. Soc.* xxxi, 143 (1931).

the country as a whole. Generally speaking, these variations can be summed up by saying that the period 1815 to 1837 was marked by depression; that of 1838 to 1874 was marked by improvement and prosperity; while after 1874 the century was again characterised by adversity and difficulty. It is in the light of this general curve that the evidence for Cambridgeshire must be examined.

The County shared with the rest of England in the disastrous effects of the Napoleonic wars. The year 1815 brought peace and beggary. Between 1814 and 1816, agriculture passed suddenly from prosperity to extreme depression. As Richard Preston asked, "Was Great Britain ever before in so reduced and impoverished a condition?"[1] As for Cambridgeshire, Lord Brougham, speaking in the House of Commons on 9 April 1816, said:

> The petition from Cambridgeshire presented at an early part of this evening, has laid before you a fact to which all the former expositions of distress afforded no parallel, that in one parish, every proprietor and tenant being ruined with a single exception, the whole poor-rates of the parish thus wholly inhabited by paupers, are now paid by an individual whose fortune, once ample, is thus entirely swept away.[2]

In the same year, it was said that "a detestable spirit of conspiracy" was manifesting itself "in the counties of Norfolk, Suffolk, Huntingdon and Cambridge, directed against houses, barns and rick-yards, which were devoted to the flames".[3] This was generally ascribed to a "want of agricultural employment, joined to the love of plunder".[3] In some localities, the general unrest broke out into riots, and the number of labourers, committed to the county gaol under the Vagrancy Law for "refusing to work for the customary wages", rapidly increased from the twenties onwards.[4]

In December 1829, came a petition from the farmers of Ely to Parliament. It could but repeat what was well known already. The labourers, "no longer able to maintain themselves by the sweat of their brows", were driven "to the scanty pittance derived from the parish funds".[5] Frequently, their distress sought a violent outlet. There was an outbreak of

[1] Richard Preston, "Review of the Present Ruined Condition of the Agricultural and Landed Interests", *Pamphleteer*, vii, 150 (1816).

[2] *Speeches of Henry, Lord Brougham*, i, 504 (1838). Lord Ernle notes that, in 1815, nineteen farms in the Isle of Ely were without tenants; and that the number of arrests and executions for debt in the Isle increased from 57 in 1812–13 to 263 in 1814–15 (*English Farming Past and Present* (1932), pp. 322–3).

[3] *Annual Register* (1816), p. iv.

[4] See E. M. Hampson, *The Treatment of Poverty in Cambridgeshire*, 1597–1834 (1934), p. 196.

[5] See *ibid.* p. 215.

rick-burning in Cambridgeshire, as in England generally. The commissioners appointed to investigate the causes of these disturbances found "distress and want of employment" all through the County.[1]

Of course, all years were not equally bad, and amidst many variations of statement and opinion, it is difficult to assess the degree of distress at any particular moment.[2] But the weight of Mr Thurnall's evidence, in 1836, leaves no doubt about the general picture.[3] Land in a neglected state was "every day increasing in quantity" owing to the "low price of agricultural produce".[4] On the other hand, he thought that, as yet, no land had been "thrown out of cultivation". Still, the condition of the tenantry was "verging on insolvency".[5] Rick-burning was frequent, and several of his best and honest labourers were threatening to rob on the highway before they would "go to the union work house". They were "ripe for everything in the world", ready to be stirred into "a state of revolution".

The accession of Queen Victoria in 1837 coincided with the beginnings of improvement. The formation of the Royal Agricultural Society in the following year was at once a symptom of revival and an aid to prosperity. Despite ups and downs, the decade that followed was marked by an advancement that reflected itself in one of England's most agricultural of counties. In his survey of 1847, Samuel Jonas declared that "few counties, if any, have improved more in cultivation than Cambridgeshire has lately done".[6]

> All the open common-fields have been enclosed (with the exception of five or six parishes), and instead of a system of cropping so exhausting to the land as a fallow and two white-straw crops in succession, with other men's flocks of sheep eating up your food and preventing improvement, we now see the land farmer on the four course system—the best that can be adopted, unless on very fine land.

[1] *Parliamentary Papers* (1834), xxxiv, Appendix B, Pt. v, to the Report on Poor Laws, pp. 49–72.

[2] Thus a calculation of the amount of unemployment in Cambridgeshire in 1830–1831 stated that "the total number of unemployed labourers in 156 parishes [out of a total of 164] in Cambridgeshire was 811; not one sixteenth of the total number of labourers, very little more than five men being so reckoned to a parish, and one man to a population of 169". And, again, "we cannot suppose any to remain unemployed during the three months which hay and corn harvest last". *Parliamentary Papers* (1834), xxxvii, Appendix C to the Report on Poor Laws, pp. 72–3.

[3] *Rep. Select Committee on Agricultural Distress* (1836), viii, Pt. 1, pp. 115 *et seq.*

[4] He attributed this to the contraction of the currency; "that is the main cause; Irish produce is another cause; want of protection against foreign corn is another; but I should say that the contraction of the currency is the main cause". (*Ibid.* p. 121.)

[5] For this, and the remaining quotations in the paragraph, see *Rep. S.C. Agricultural Distress* (1837), v, 129.

[6] For the quotations that follow, see S. Jonas, "On the Farming of Cambridgeshire", *Jour. Roy. Agric. Soc.* (1847), p. 35.

Large flocks of sheep were fattened with corn and cake for the London markets; indeed, Mr Jonas Webb of Babraham was "one of the first and most justly celebrated breeders of Southdown sheep in existence". Large numbers of cattle were also to be seen. "Comparing the present system with the former," wrote Jonas, "it is astonishing to mark the increased wealth our present improved system brings to the state; not only thus largely increasing the national wealth, but also giving full employment for our labourers."

Jonas divided Cambridgeshire into four districts:

(1) "The southern and central part of the county, extending from Ickleton to the north side of Newmarket, is light land, consisting of chalk, sands, tender loams, and gravels." On these "thin-skinned, poor, light, hungry lands", where turnips formed part of the rotation, the application of bones and guano had done much; at Duxford and Whittlesford were "two very extensive and most excellent bone-mills".

(2) The eastern side of the county, adjoining parts of the counties of Essex and Suffolk, up to Cheveley, near Newmarket, was heavy clayland of various qualities, "all well hollow-drained, and generally speaking well farmed".

(3) Thirdly, came "the Fen district, an accumulation of vegetable deposit resting on the fen-clay". The improvements in this district due to draining and claying the land were "truly wonderful. Drainage condenses the land, and claying consolidates it".[1]

(4) Lastly, "the western side of the county, adjoining Bedfordshire, Hertfordshire, and Huntingdonshire, consists of a tough tenacious clay of little value on the hills,[2] but the flats are good, strong, deep, staple lands". This area was not as well managed as the eastern clayland, "particularly as relates to draining".

But although, generally speaking, "improvements and high farming" were bringing prosperity to the County, Jonas had to confess that "there yet remained some districts that were badly cultivated". He seems to have had the western clays particularly in mind.

The picture that James Caird gave of the County in 1850–51 is quite another impression:

In any district of England in which we have yet been, we have not heard the farmers speak in a tone of greater discouragement than here. Their wheat crop, last year, was of inferior quality, the price unusually low, and to add to this, their live stock and crop are continually exposed to the match of the prowling incendiary.[3]

[1] For the importance of claying the fen peat, see p. 121 above.
[2] See p. 27 above.
[3] J. Caird, *English Agriculture in* 1850–51 (1852), pp. 477 *et seq.*

Fires were "of almost nightly occurrence". "A few bad fellows in a district are believed to do all the mischief, and bring discredit on the whole rural population." But this does not imply that the rural population had no grievance; although employment was available, wages were low, "7s. to 8s. a week being the current rate".[1] Taken together, these two pictures, by Jonas and Caird respectively, of technical improvement and social discontent, may give some idea of conditions in Cambridgeshire during the middle of the century.

However conflicting the evidence, nothing can gainsay the fact that many parts of rural Cambridgeshire had a special boom of their own, after about 1850, through the setting up of the "coprolite" industry for the manufacture of manures. By origin, the word "coprolite" signifies petrified dung, presumably of enormous reptiles, but the term came to include phosphatised casts of vertebrate remains in general. Coprolites were to be found in the Cambridge Greensand[2] that marked the base of the Chalk, and that ran north-east—south-west through Soham, Burwell, Swaffham, Horningsea, Cambridge, Grantchester, Barrington, and so westward into Bedfordshire.[3] The deposits were described by O. Fisher in 1873:

> The Cambridgeshire phosphatic nodules, as is well known, are extracted by washing from a stratum (seldom much exceeding a foot in thickness) lying at the base of the lower chalk, and resting immediately, without any passage-bed, upon the Gault. There is, however, a gradual passage upwards from the nodule-bed into the lower chalk or clunch. The average yield is about 300 tons per acre; and the nodules are worth about 50 shillings a ton. The diggers usually pay about £140 per acre for the privilege of digging, and return the land at the end of two years properly levelled and re-soiled. They follow the nodules to a depth of about 20 feet; but it scarcely pays to extract them to that depth.[4]

Generally speaking, the years between 1850 and about 1870 were prosperous ones for much of the County. A footnote in the Census Returns attributes an increase of population at Orwell, between 1861 and 1871, to the "demand for labour in the coprolite diggings". The same cause, too, was responsible for growth at Barton, Great Eversden, Harston, Haslingfield and Trumpington; and at Wicken the increase was likewise "attributed to the extensive coprolite digging having attracted numbers of labour". By the end of the century, however, the coprolite beds had

[1] J. Caird, *op. cit.* p. 468.

[2] See p. 13 above.

[3] The Lower Greensand phosphatic deposits were also being worked about the years 1866–68, chiefly near Wicken. W. Keeping, *Fossils of the Neocomian Deposits of Upware and Brickhill* (1883), pp. 1–2. See p. 11 above.

[4] O. Fisher, "On the Phosphatic Nodules of the Cretaceous Rocks of Cambridgeshire", *Quart. Jour. Geol. Soc.* xxix, 52 (1873). A detailed account of coprolite digging, based upon direct observation at Burwell, is given by C. Lucas, *The Fenman's World* (1930), p. 25.

become practically exhausted,[1] and were only temporarily revived during the Great War of 1914–18.

Despite the coprolite prosperity, Cambridgeshire shared in the general ebb that marked English agriculture from the seventies onwards. An idea of the nature of farming in the County is provided by the following figures,[2] derived from the Agricultural Returns of 1874:

| | Cambridge | England |
|---|---|---|
| Percentage of corn crops to cultivated land | 53·4 | 31·4 |
| Number of cattle per 100 acres | 9·9 | 17·0 |
| Number of sheep per 100 acres | 67·4 | 80·0 |

The type of farming indicated by these figures made the district very susceptible to the depression that started[3] between 1875 and 1879. Mr Druce, who visited the County in 1880, attributed the depression, first and foremost,

to a succession of four or five years' deficient harvests [due to wet seasons], accompanied with extremely low prices, occasioned by the excessive importations from America. Among the contributory causes were increased rates of wages and the difficulty of obtaining juvenile labour due to the Education Acts.[4]

In those districts of the County where attention had been directed to the production of meat, the depression was not so much felt; but the tenor of Mr Druce's report leaves no doubt about the general situation. In the Fenland, "one-half the farmers were absolutely insolvent, and the other half greatly reduced in circumstances". On the upland, "there were considerable quantities of land unlet and which could not be let".

Nor did the story end there. In another report, made in the following year, Mr Druce found "that the depression has very much increased";[5] and he added that the Agricultural Returns for 1881 "afford confirmatory evidence of the continuance and depth of the depression in this county". The numbers of stock were continuing to decrease, notwithstanding an earlier decrease since 1875. Mr Druce graded the intensity of the depression among the counties that comprised his district as: (1) Huntingdon, (2) Essex, (3) Cambridge.

[1] E. Conybeare, *A History of Cambridgeshire* (1897), p. 269. See also T. M. Hughes and M. C. Hughes, *Cambridgeshire* (1909), p. 112.

[2] F. Clifford, *The Agricultural Lock-Out of* 1874 (1875), pp. 339–40.

[3] *Royal Commission on Agriculture: Report on Cambridgeshire, Report by Mr Wilson Fox* (1895), p. 25.

[4] *Royal Commission on Agriculture, Report by Mr Druce* (1881), p. 365.

[5] *Royal Commission on Agriculture, Report by Mr Druce* (1882), pp. 14–20.

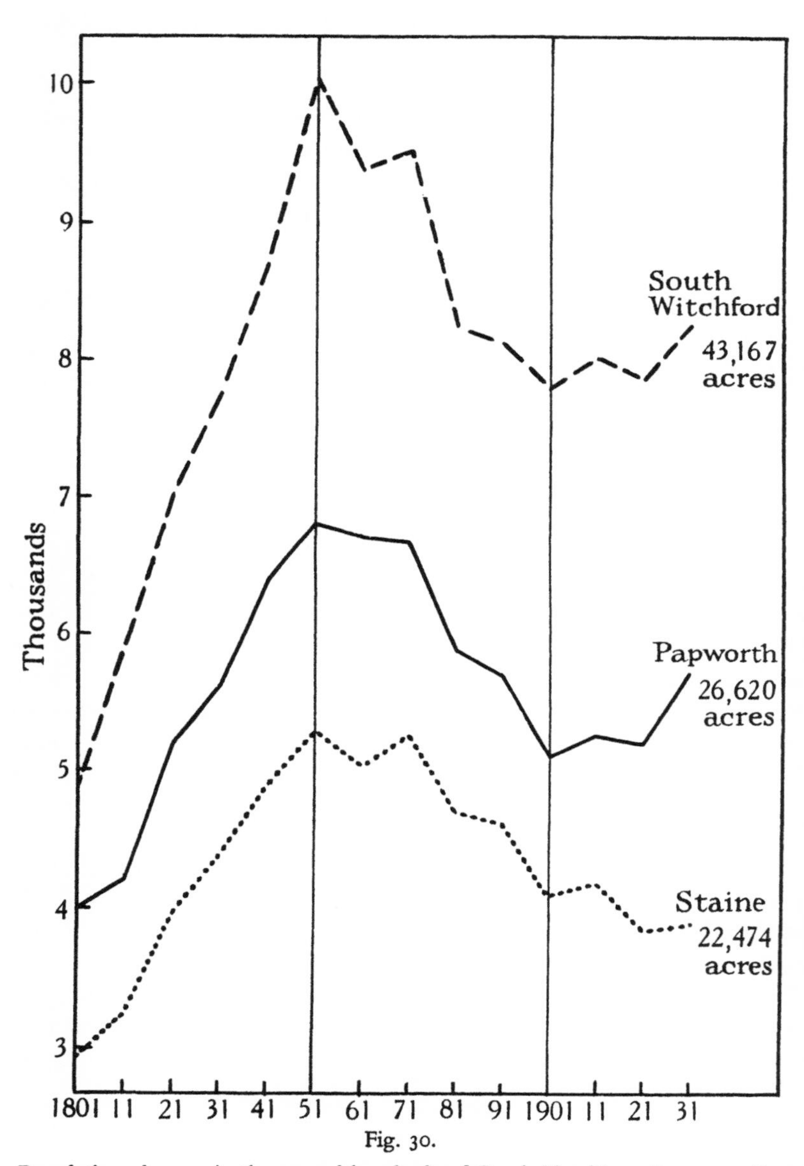


Fig. 30.

Population changes in three rural hundreds of Cambridgeshire, 1801–1931. I am indebted to the Editor of the *Victoria County Histories* (Mr L. F. Salzman) for access to the Population Tables (by G. S. Minchin) in the forthcoming *Cambridgeshire*, vol. ii.

This second half of the nineteenth century was also marked by another feature in the rural circumstances of the County. During the earlier half of the century, population had continued to grow despite distress and unemployment. But before the middle of the century, this situation was changing. A hint of things to come is provided by that footnote, in the 1841 Census Returns, which states that, from Willingham, "upwards of 100 persons have emigrated to the United States since 1831". At Wimpole, a decrease was "attributed to several large families having left the Parish, and others having emigrated since 1831". The 1851 Returns noted that the decrease at Croxton was also due in part to emigration, as was that at Wimpole and West Wratting. The 1861 Returns have very many of these references. One footnote tells its own story:

> General decrease of population throughout the district of Caxton and especially in the parish of Caldecote is mainly attributed to emigration and migration owing to lowness of wages, etc.

Similar causes helped to account for a decrease in thirty other villages in the County. The reason stated was sometimes "emigration"; sometimes "migration of labourers to towns", or to "manufacturing districts", or to "London and the north of England", or to "Manchester and its vicinity", or, again, to "the metropolis and other large towns". It is true that the Census footnotes also record some increase due to "the erection of new cottages on a recent enclosure", as at Gamlingay and Hardwicke; or due to a temporary influx of labour employed upon railway construction at Great Shelford and Harston, or employed upon a new cut at Clenchwarton.[1] At Sawston and Whittlesford, the increase was "due to the paper mill and parchment factory at Sawston". Then, too, there were the attractions of the coprolite diggings[2]; there were also some miscellaneous explanations.

After 1871, the explanatory footnotes cease to appear in the Census Returns, but the figures themselves tell their own story. Fig. 30 sums up the evidence for three rural hundreds in the County, and shows quite clearly how the countryside was emptying itself. The difference between this diagram and that of Fig. 31 is explained by the growth of Cambridge,[3] Wisbech, Ely, March, and Whittlesey, and also by local circumstances (e.g. jam-making at Histon). By the end of the century, the urban and semi-urban centres had grown; the rural settlements had become smaller.

A full picture of rural conditions towards the end of the century is given in the Cambridgeshire section of the *Report of the Royal Commission on Agriculture* (1895). This drew a great distinction between north and south Cambridgeshire:

[1] The Eau Brink Cut; see pp. 120, 191. [2] See p. 126 above. [3] See Fig. 41.

In the north there are a number of districts where the fen land has depreciated but little, and some where it has not depreciated at all, while in the south there are large tracts where the deteriorated state of the land is painfully apparent to all, being practically worthless to owner and occupier alike, and scarcely able to be designated as cultivated. Between Cambridge and Huntingdon the state of the land is as bad as in the worst districts in Suffolk, and in some other localities it is little if any better.[1]

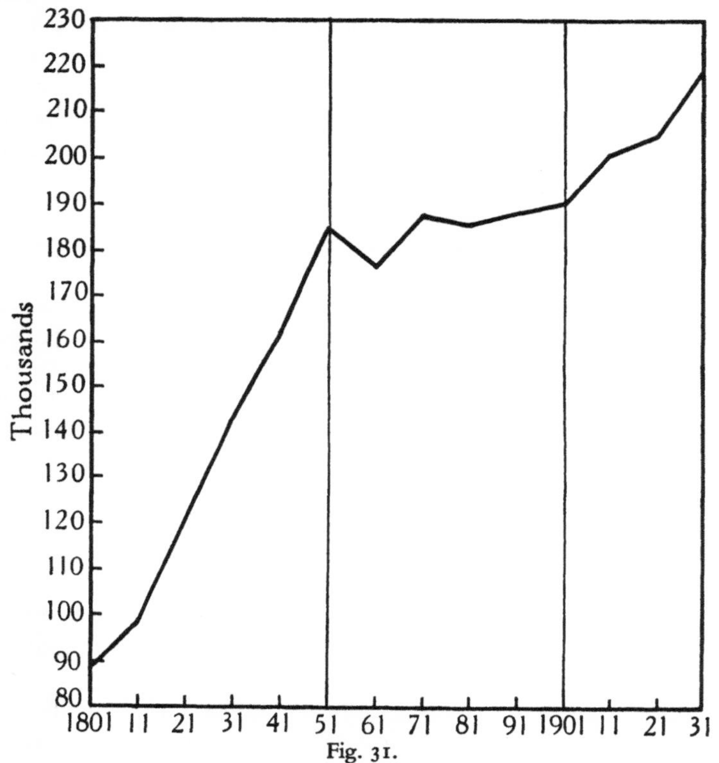


Fig. 31.

Population changes in Cambridgeshire (including the Isle of Ely), 1801–1931. I am indebted to the Editor of the *Victoria County Histories* (Mr L. F. Salzman) for access to the Population Tables (by G. S. Minchin) in the forthcoming *Cambridgeshire*, vol. ii.

On the upland areas of the County, it was generally acknowledged that there was "a great deal of rough land very nearly out of cultivation". Captain Hurrell of Madingley, "in a nine mile run with the hounds",[2]

[1] W. Fox, *Royal Commission on Agriculture: Report on Cambridgeshire* (1895), pp. 25–6. "Between Cambridge and Huntingdon" was heavy clayland (see Fig. 29). See also R. Bruce, "Typical Farms of East Anglia", *Jour. Roy. Agric. Soc.* (1894), p. 497, for details of farms at Barton, Bourn, Linton, Little Eversden, Littleport, Trumpington, and Whittlesford.

[2] See W. Fox, *op. cit.* pp. 26–7 for the quotations that follow in this and the next paragraph.

rode over only nine arable fields; "most of it had been seeded down". On "the boulder clay formation to the west of Cambridge, a considerable area" had been left uncultivated.[1] There were also "much fewer stock and sheep being kept in the county". Mr Dymock, who farmed 600 acres at Waterbeach, said that the condition of the land had been going back for twelve years. "It began in the bad season of 1879, when the heavy land got into a very bad state. Then bad prices came, and hence so much money could not be spent on it." Mr W. J. Clark, of Thriplow, could "point to farms that 10 years ago were patterns for cleanliness and good farming that are now in a deplorable state". Arrears and reductions of rent were "undoubtedly large in number".[2] Mr Martin Slater, of Weston Colville, thought that the land had "very greatly gone back in condition during the last 25 years in his district". Of the land outside the Fens, "the turnip and barley land near Newmarket" (i.e. light land) was said to have suffered least.

The evidence from the Fenland was less doleful. It was true that some localities had deteriorated, "partly from the effect of the seasons and partly from want of capital". Since the depression, fewer cattle and sheep had been kept. At Chatteris, it was stated that "the high lands and gravel lands have certainly gone back". That all was not desolation, however, can be seen from the following statement made at a meeting of farmers at Wisbech in 1894:

Generally speaking, the strong land has deteriorated. The wet seasons had a great deal to do with it, as well as loss of capital. Last year [1893] did a lot to help the strong land. Men will not put money into strong land farming. The acreage of wheat crop has decreased by 25 per cent in this district. The fen land has gone back very little in condition; but it is not clayed so much, partly from want of capital, but partly because it is becoming stronger on account of the peat disappearing owing to the drainage.[3] The marsh land has not gone back a bit between Wisbech and Long Sutton; there has been the means of enabling the people to escape from the depression. They are able to grow the best class of potatoes, vegetables, and fruit. The men in the marsh have been hit to some extent by prices, but are better off than other people occupying land.

Thus was a new element called in to redress the balance of the older economy. The first orchard had been planted in the Wisbech area as early as the fifties; now, in the eighties and nineties, many farmers found themselves forced to adopt a fresh form of husbandry, and so turned to market gardening and fruit farming.[4] The new crops had also been spreading on the upland.[5] The Chivers' enterprise around Histon dates from the middle

[1] See p. 56 above. [2] W. Fox, *op. cit.* p. 34. [3] See p. 120 above.
[4] See C. Wright and J. F. Ward, *A Survey of the Soils and Fruit of the Wisbech Area* (1929), pp. 25–7.
[5] J. F. Ward, *West Cambridgeshire Fruit-Growing Area* (1933), pp. 29–33.

of the century;[1] while, at Rampton and Cottenham, a considerable amount of fruit was being grown by the villagers. In 1873, there were about 1000 acres of fruit within 10 miles of Histon; by 1894, this acreage had increased to 3000. The other fruit-farming area on the upland was around Meldreth and Melbourn, where, during the fifties, a substantial acreage of fruit had been planted.

It is little wonder, then, that the *Report* of 1895 could state that the profits made from fruit growing and market gardening "have in the last few years been more satisfactory than those from ordinary farming".[2] From the depression of the nineteenth century, the twentieth was to inherit at any rate some beginnings of prosperity.

## NOTE ON RAILWAY CONSTRUCTION

"The principal rivers are the Cam or Granta, and the Ouse: the latter river is navigable from Cambridge to Lynn, in Norfolk, to which port large quantities of the grain produce of this county hitherto have been sent by this navigation; but it will soon be a question whether the corn-produce will not in future travel to London by the railroad".[3] Thus wrote S. Jonas in 1847, two years after the opening of the London-Cambridge-Norwich main line. The other railway lines quickly followed.

The lines passing through the County mostly formed part of the Great Eastern system. But three other railway companies also ran over lines of the G.E.R. Co. to Cambridge: the Great Northern from Hitchin via Shepreth; the Midland from Kettering via Huntingdon; and the London and North-Western from Bedford and Bletchley via Hills Rd. Junct. Cambridge; while, in the north of the County, the Peterborough, Wisbech, Sutton Railway was part of the Midland and Great Northern Joint Committee's line. The various lines (see Fig. 32) were opened[4] at the following dates:

| | |
|---|---|
| 1. The G.E.R. main line from London to Norwich, entering the County at Chesterford, and leaving it after passing through Cambridge and Ely. | 30 July 1845. |
| 2. Ely to March and Peterborough. | 9 December 1846. |
| 3. March to Wisbech. | 3 May 1847. |

[1] H. Rider Haggard, *Rural England* (1902), ii, 51. In 1873, the manufacture of jam was started at Histon. See p. 156 below.

[2] W. Fox, *op. cit.* p. 6.

[3] S. Jonas, "On the Farming of Cambridgeshire", *Jour. Roy. Agric. Soc.* (1847), p. 38.

[4] For this information I am much indebted to Mr J. H. Wardley of King's Cross Station. Mr E. D. Robinson of Cambridge has also given me help in this connection: an older list is in E. Conybeare's *A History of Cambridgeshire* (1897), p. 279.

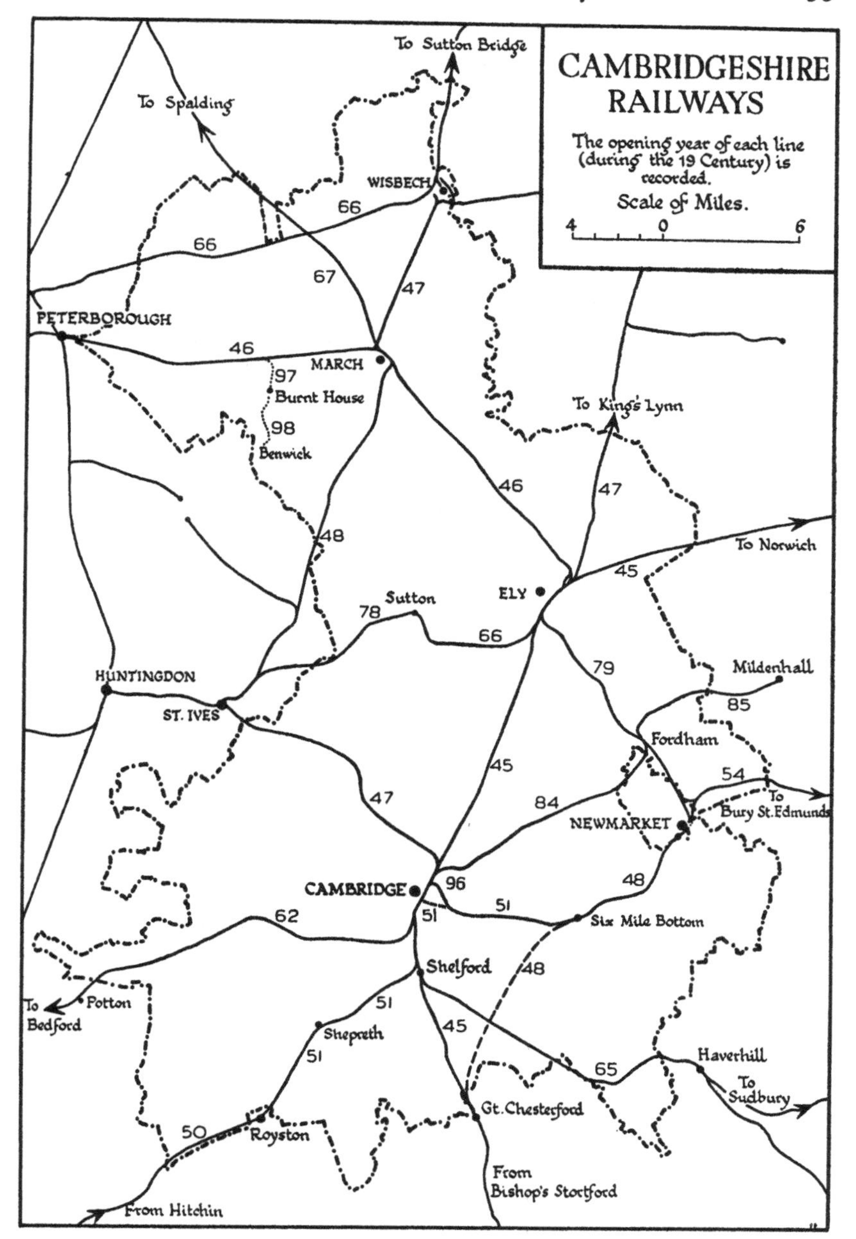


Fig. 32.

I am indebted to Mr J. H. Wardley of King's Cross Station for the information on this map.

| | | |
|---|---|---|
| 4. | Cambridge to St Ives and Huntingdon.[1] | 17 August 1847. |
| 5. | Ely to Lynn. | 26 October 1847. |
| 6. | March to St Ives. | 1 February 1848. |
| 7. | (*a*) Chesterford to Newmarket.[2] | 4 April 1848. |
| | (*b*) Newmarket to Bury St Edmunds. | 1 April 1854. |
| 8. | (*a*) Hitchin to Royston. | 2 October 1850. |
| | (*b*) Royston to Shepreth. | 3 August 1851. |
| 9. | Shepreth to Shelford.[3] | 25 April 1851. |
| 10. | Cambridge to Six Mile Bottom.[4] | 9 October 1851. |
| 11. | Bedford to Cambridge (L.N.W.R.), entering the County north of Potton. | 1 August 1862. |
| 12. | Shelford to Haverhill (Suffolk). | 1 June 1865. |
| 13. | March to Spalding.[5] | 1 April 1867. |
| 14. | (*a*) Ely, Haddenham and Sutton. | 6 April 1866. |
| | (*b*) Sutton to Needingworth (Hunts). | 10 May 1878. |
| 15. | Peterborough, Wisbech, and Sutton. | 1 August 1866. |
| 16. | Ely to Newmarket. | 1 September 1879. |
| 17. | (*a*) Cambridge (Barnwell) to Fordham. | 2 June 1884. |
| | (*b*) Fordham to Mildenhall. | 1 April 1885. |
| 18. | (*a*) Goods line from Three Horse Shoes Junction to Burnt House Siding. | 1 September 1897. |
| | (*b*) Burnt House Siding to Benwick. | 2 August 1898. |

[1] By agreement of 26 June 1864, the Midland trains ran from Kettering to Cambridge over this line.

[2] The section of this line from Chesterford to Six Mile Bottom (about 12 miles in length) was closed on 9 October 1851, upon the opening of the line from Six Mile Bottom to Cambridge. It was abandoned by the Eastern Counties Railway Act of 1858. The deserted cuttings and embankments are still striking features of the landscape.

[3] The G.N.R. were compelled by their Act to permit the G.E.R. to meet them at Shepreth, and did not get running powers over the line to Cambridge until 1866. Before the Shelford and Shepreth line was made available, the G.N.R. used to run coaches from Shepreth to Cambridge by road, in connection with their trains, timed to do the distance (9 miles) in 40 minutes.

[4] To join the unabandoned section of the Chesterford-Newmarket line. There was an extension from Newmarket to Bury St Edmunds on 1 April 1854. The junction at Cambridge Station was taken out when the present diversion line over Coldham's Common was opened on 17 May 1896.

[5] Originally a G.N.R. line, but owned jointly with the G.E.R. up to the Railways Act of 1921.

CHAPTER TEN

# THE AGRICULTURE OF CAMBRIDGESHIRE

By R. McG. Carslaw and J. A. McMillan

## (A) THE PERIOD 1900–1936

By R. McG. Carslaw, M.A., PH.D.

THE FOUR PHASES EVIDENT IN THE FARMING OF THE COUNTRY since the beginning of the century have been well marked in Cambridgeshire:

(i) The pre-war years up to 1914, when, on the whole, profits and wages were gradually rising. During this time, adjustments in cropping, in livestock policies, and in methods of production, were being methodically, if slowly, evolved.

(ii) The abnormal war, and immediately post-war, years of 1914–20, characterised by scarcity prices, by high profits, and by the improvisation of methods to meet a shortage of labour and raw materials.

(iii) The post-war depression of 1921–31, with heavy capital losses, with statutory minimum-wage legislation, and with much searching for new methods and types of farming, e.g. the development of sugar-beet growing, poultry, motor tractors, etc.

(iv) The years 1932–36, marked by the combined effect of (*a*) Governmental action, e.g. subsidies, tariffs, and quotas; (*b*) Marketing Boards (milk, pigs, potatoes, etc.); (*c*) cheap feeding stuffs; and (*d*) improved technical efficiency. This period, too, has been characterised by better profits and by rising wages.

Between 1900 and 1936, there was a decrease of over 20,000 acres (about 4 per cent) in the area under crops and grass (from 490,306 acres in 1900, to 467,980 acres in 1936).[1] Rather more than one-half this decline can be attributed to the deterioration of cultivated land, particularly since 1920, into "rough grazings"; but as much as 10,000 acres was lost to agriculture as a result of the encroachment of buildings, roads, etc. In spite of this decline in cultivated area, there is reason to believe that the

[1] The figures in sections A and B of this chapter are derived largely from Ministry of Agriculture Statistics.

aggregate agricultural production has not diminished. The principal "cash crops" for farmers in the County are wheat, barley, sugar beet, and potatoes. As the following table shows, the combined area of these crops has increased considerably:

| | 1900 | 1910 | 1920 | 1930 | 1936 |
|---|---|---|---|---|---|
| | acres | acres | acres | acres | acres |
| Wheat | 95439 | 100432 | 96091 | 81552 | 104790 |
| Barley | 52484 | 53821 | 56071 | 47703 | 32995 |
| Potatoes | 22790 | 26865 | 36416 | 32152 | 41324 |
| Sugar beet | — | — | —* | 43970 | 41458 |
| Total | 170713 | 181118 | 188578 | 205377 | 220567 |

* Less than 50 acres were grown in 1919.

The increased acreage devoted to "cash crops" has been secured by reducing the area under crops grown primarily for fodder—turnips, swedes, mangolds, oats, and rotational grasses. As the table below shows, there has been a steady decline from 1900 to 1936 in these principal fodder crops:

| | 1900 | 1910 | 1920 | 1930 | 1936 |
|---|---|---|---|---|---|
| | acres | acres | acres | acres | acres |
| Turnips and Swedes | 15755 | 14114 | 9461 | 5868 | 2830 |
| Mangolds | 17858 | 16934 | 15184 | 9336 | 6952 |
| Oats | 49619 | 47220 | 46303 | 41456 | 33540 |
| Clover and rotational grasses | 57166 | 44305 | 40813 | 38770 | 28746 |
| Total | 140398 | 122573 | 111761 | 95430 | 72068 |

This very startling reduction in the acreage of fodder crops has been accompanied by a large reduction in the number of sheep and a marked decrease in the cattle population:

| Live stock | 1900 | 1910 | 1920 | 1930 | 1936 |
|---|---|---|---|---|---|
| | number | number | number | number | number |
| Cows and heifers in milk and in calf | 17835 | 17206 | 16610 | 16434 | 18704 |
| Other cattle | 39277 | 41873 | 30459 | 29235 | 26631 |
| Sheep | 208272 | 168778 | 73543 | 71926 | 62548 |
| Pigs | 46180 | 51959 | 52021 | 72653 | 118051 |
| Poultry | ?* | 564794† | 508534‡ | 791637 | 926716 |
| Horses for agriculture | 21808 | 23609 | 19987 | 18552 | 15461 |

* Not known. † 1913. ‡ 1921.

On the other hand there have been increases in pigs and poultry, types of live stock for which special fodder crops are seldom grown, and which are primarily dependent on concentrated feeding stuffs. The crude figures in the annual 4th June statistics show, between 1900 and 1936, decreases of 12,000 in cattle, and 146,000 in sheep, and increases of 72,000[1] in pigs, and possibly 400,000 in poultry. Though it is very difficult to reduce the different categories of live stock to a common denominator, it seems probable, on balance, that the monetary value of the livestock output may even have increased during the period.

The explanation of this apparent anomaly is naturally complex. Undoubtedly, farmers have become increasingly dependent upon purchased feeding stuffs for their live stock, particularly after 1930. The expansion in livestock commitments took place chiefly in pigs and poultry—two categories primarily dependent on concentrated feeding stuffs. The decline in livestock numbers has been in sheep and beef cattle, which in arable districts commonly consume large quantities of home-grown bulky foods. The number of sheep has fallen by two-thirds, and "other cattle" by one-third, as compared with a decrease of 50 per cent in the acreage of fodder crops. Undoubtedly the development of sugar beet has contributed, particularly in the case of sheep, to the maintenance of the fodder supply, for the "tops" have replaced large acreages of "sheep keep" (e.g. turnips, kale, etc.) formerly grown to be close-folded. Further, the reduction in the number of working horses (from 23,600 in 1910 to 15,500 in 1936) must have liberated a considerable area, perhaps as much as 20,000 acres, formerly required for growing horse feed. This latter economy has, of course, been at least partly off-set by increased expenditure on machinery, oil, paraffin, etc.

Judged by money values, the crop output of the County in 1936 appears to have been substantially greater than at the beginning of the century, while the livestock output was at least not smaller. Further, the area under fruit, on holdings of one acre or more, increased from some 6000 acres in 1900 to 15,000 acres in 1936. This apparent expansion in total agricultural output was secured despite a decline in the number of workers employed, and a decrease in the number of horses used for agriculture. Statistics of employment are not available for years earlier than 1921, when the number of workers (including casuals) stood at 24,610. It seems probable that in 1900 the number was larger, but by 1930 it had fallen to 23,068; and in 1936 it stood at 21,644. Thus between 1921 and 1936 there was a decline of nearly 3000 workers (12 per cent). Output per worker

[1] With two gestation periods in the year, this figure should be approximately doubled to determine the rise in the annual pig output.

must therefore have increased very markedly during the period, partly as a result of increased mechanisation (particularly tractors), partly as a result of the alterations in the types of commodities produced, and partly also, owing to greater skill in the supervision of labour.

These changes have undoubtedly been most pronounced since the war, particularly after 1930, when legislation prevented agricultural wages from falling proportionately with the drop in commodity prices. Faced with the problem of wage rates fixed at roughly double their pre-war level, farmers were forced to devise means of increasing the output per worker.[1] Broadly speaking, the years 1920–36 probably constitute a period of unprecedented rate of change both in the internal and external organisation of farming in the County.

## (B) GENERAL SURVEY

By J. A. McMillan, B.SC.

*Organiser of Agricultural Education, Cambridgeshire County Council*

Cambridgeshire and the Isle of Ely are now separate administrative units. When it will be convenient to refer specifically to one or the other, the Administrative County of Cambridge will be termed "the County", as opposed to "the Isle of Ely". Taken together, they have an area of 553,555 acres. Of this, in the year 1937, some 466,600 acres were "under crops and grass" and 12,671 acres were "rough grazings". There are 3½ acres of arable land to every acre of grassland, a concentration only exceeded in England by the Holland Division of Lincolnshire, where the proportion is four to one. A markedly rural character is reflected also in the population figures. The total population in 1931 was 217,702, a density of 260 per square mile, which compares with a density of 690 per square mile for England and Wales as a whole. Of the total employed persons over fourteen years of age, 28 per cent were engaged in agricultural occupations, compared with 6 per cent for England and Wales.

### MAIN CROPS

"Holdings of 1 acre and upwards" returned in 1937 numbered 7257; and three-quarters of these fall in the group class from one to fifty acres; while there are many holdings of less than one acre, which are not included

[1] R. McG. Carslaw, "The Changing Organisation of Arable Farms", *Econ. Jour.* xlvii, 483 (1937). On a group of 150 farms, the physical output per worker increased by some 27 per cent between 1931 and 1936.

in the official returns. Though in some districts large farms stretch as far as the eye can see, Cambridgeshire as a whole may be regarded as a county of small farms, small-holdings, and market or cottage gardens. Of the ordinary farm crops, excluding rotation and permanent grass, the most important on an acreage basis are wheat, barley, oats, potatoes, and sugar beet, which together covered 245,000 acres in 1937, more than two-thirds of the total arable acreage. Fig. 33 shows the acreage of each of these crops in 1913 and from 1919 to 1937.

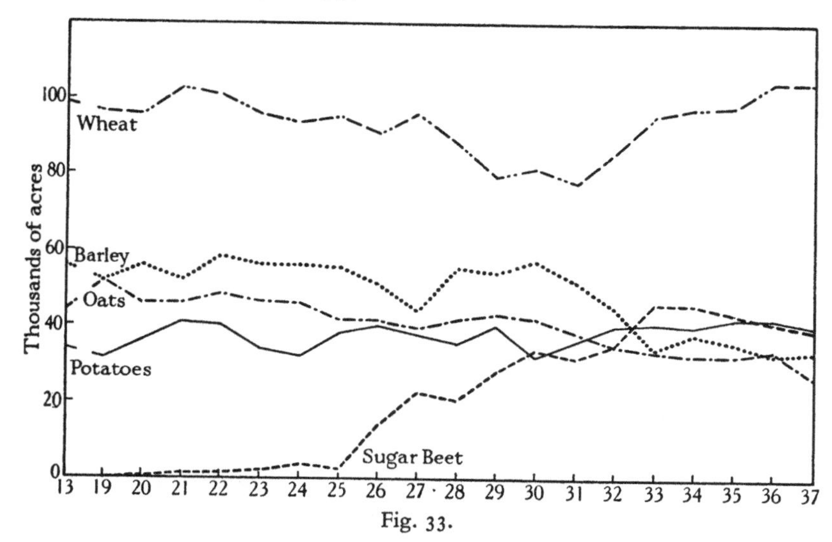


Fig. 33.

Acreages of Main Crops, 1913–37 (Ministry of Agriculture Statistics).

The *Wheat* acreage has been maintained fairly well in recent times, except during a period of low prices just before 1932. The effect of the Wheat Act of that year was to check the decline, and later to increase the acreage to a little above the pre-war level. The wide variations in soil type lead to an equally wide variation in the choice of seed. Some of the more common varieties are Little Joss, Squarehead's Master, Victor, Wilhelmina, Yeoman, and Rivett's. In the County, a distinctive feature is the large proportion of wheat which is grown after a one year's sainfoin or clover leyer. If the latter has not been heavily folded with sheep, it is customary to apply a dressing of farmyard manure prior to ploughing for the wheat crop. In the Fenland wheat usually follows a fallow crop and receives no special manuring in the autumn.

The *Barley* acreage has been declining gradually in recent years. In the County, the acreage has been fairly constant between 30,000 and 37,000 acres. In the Isle, however, where there are greater difficulties in growing a good malting sample, the area under barley is now only one-quarter of the pre-war figure. On the lighter and better barley soils it is indeed rare to find a field sown with any other variety but Spratt Archer. On the heavier soils and in the Fens, some choose Plumage Archer.

The *Oat* acreage is distributed fairly evenly between the County and the Isle. In the former, the greater part of the oats is autumn-sown, and Grey Winter is a popular choice. Though inclined to "lodge" at times, this variety proves a reliable cropper, and is liked by those who buy for the racing stables at Newmarket. Marvellous and Resistance, often sown in the very early spring, are also widely grown. Spring oats, when sown in March, crop reasonably well as a rule, and are grown on a limited area. Victory is the variety most in favour.

*Potatoes.* Rather more than nine-tenths of the 40,000 acres of potatoes are grown in the Isle (see Fig. 34), where this is one of the principal crops contributing to farm income. The tendency in recent years has been to concentrate potatoes on those soils proved to be best suited to the crop, and to manure more intensively than formerly. Now, too, only a few proved varieties are grown at all widely; recent reports of the Potato Marketing Board indicate that well over 30,000 acres are planted with Majestic and King Edward VII. The cultivation of early varieties is limited to some 3000 acres on the lighter and more silty soils, Eclipse being the most commonly grown.

The *Sugar-beet* acreage has increased very considerably since the thirty-nine acres that were grown in 1919. Some two-thirds of the total acreage is now grown on the richer fen soils (see Fig. 35), where yields considerably above the average for the country are obtained in most seasons. It is rather remarkable how this new crop has taken its place in the farm rotation without any considerable upheaval in farming practice. It has replaced fodder crops, rotation grasses, mustard for seed, oats, and barley, the latter particularly in the Isle. Its introduction has resulted in busier times in early summer and late autumn. On those farms where it is grown on any extensive scale, it is necessary to employ casual labour to assist the regular farm staff.

*Leyers* are to be found almost exclusively in the County, where they are a definite feature of the light and heavy land rotations. On chalk soils, a one year's ley of sainfoin or broad red clover is extremely common, the crops being either folded and seeded, hayed and seeded, or merely folded. Grass and clover mixtures are not now so popular, partly on account of the smaller demand and lower prices for this type of hay in recent years, and partly because the crops which follow appear to yield less well after mixtures. On both light and heavy soils, the leyers are ploughed up after one year, except on certain of the latter soils where there is a growing appreciation of the value of the system of alternate husbandry. The failure of the leyer on the lighter soils may well prejudice the yields of the other crops in the rotation and in consequence great importance is attached to its proper establishment and careful management.

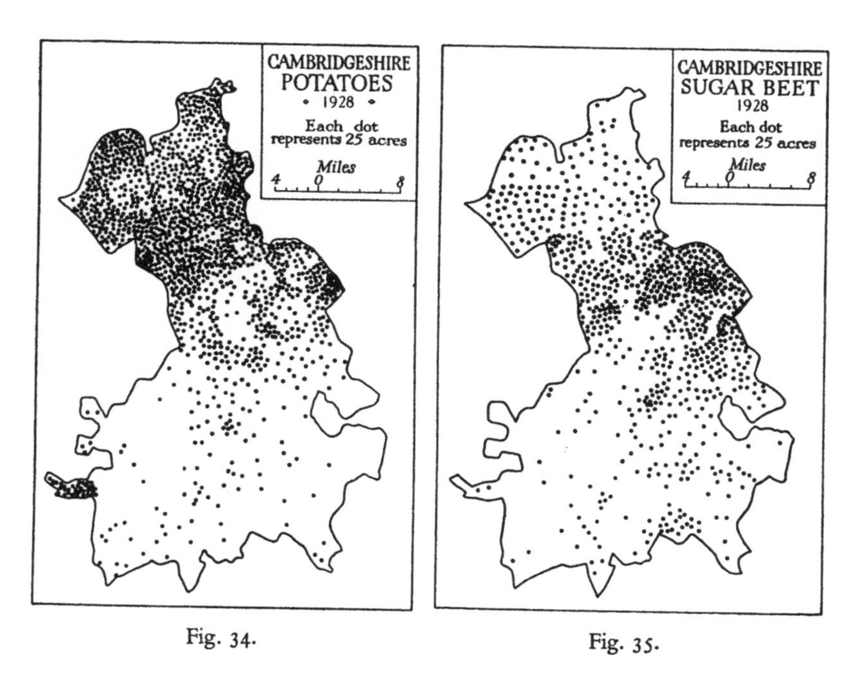


Fig. 34.

Fig. 35.

Redrawn from M. Messer, *An Agricultural Atlas of England and Wales* (1932).

*Mustard* for seed is also an important crop. It may take the place of a cereal or fallow crop, and in a catchy season the land upon which it is to be sown may not be determined until late spring. The heavier yields are obtained on the fen soils, but excellent results are also recorded on the lighter and heavier soils, which carry some 50 per cent of the 8000 odd acres grown each year. A further considerable acreage is sown each year with mustard for folding or for ploughing in.

*Beans*, long regarded as a standard crop of a heavy land rotation, are not now entitled to a place of prominence. The 6000 acres grown in 1937 represented only one-quarter of the area under this crop in 1913. There is little doubt that the change has arisen from a desire to grow crops giving a higher gross return per acre than can normally be obtained from the bean crop.

*Mangolds* and the various types of Brassicae for sheep feed are grown to a less extent each year. The combined acreage under these crops in 1937 was only 12,000 acres, the lowest return of post-war years. Though the larger decline in 1937 may in part be due to adverse weather conditions, there is little doubt that the establishment of the sugar-beet crop and the reduction in the number of arable flocks of sheep and of winter-fed bullocks have played the major part in the gradual reduction of the acreage under these crops.

*Market-Garden Crops* occupy a relatively small acreage, but make no mean contribution to the gross income per acre in those areas specially selected for their cultivation. With the exception of crops such as asparagus, and of very limited areas (e.g. on the Gamlingay Greensand), market-garden crops are taken generally in the ordinary farm rotation. The two principal crops of this nature, celery and carrots, are grown chiefly in the Isle. Celery thrives well in the cool, deep and moist black fen soils and its cultivation now covers over 3000 acres—three times the area in 1913 and more than one-third of the total celery acreage in England and Wales as a whole. Carrots, chiefly of the stump-rooted type, are grown extensively in certain well-defined areas, e.g. around Chatteris. Other crops of some importance are peas and beans for pulling, grown chiefly in the Isle, Brussels sprouts (grown mainly on some of the stiffer soils in the County), cabbages, cauliflower, broccoli, and asparagus, though the latter has not been grown so widely of late years.

It will have become evident that the rotations to accommodate the large number of crops already mentioned must show some considerable variation. In general they vary from three to six courses. On the one hand, a three-course system of two fallow crops and a corn crop is common on the best fen soils. At the other extreme, there is the light land three- or six-course rotation of the mechanised farmer, who hopes to grow corn on two-thirds of his arable acreage each year. Then there is the standard four-course rotation of the heavy-land farmer, where 50 per cent of the land is cropped with corn; and the common rotation of the chalk farm, viz. fallow crop, corn, corn, seeds, corn, though in this case there are modifications in the arrangement of the crops and some still prefer the Norfolk four-course rotation.

*Fruit growing and flower culture* are concentrated in certain well-defined areas, which were mainly under grass, or part of an ordinary mixed farm rotation, until some sixty years ago.[1] With certain notable exceptions, fruit

[1] See pp. 131 and 153.

and flower growing are in the hands of small-holders, quite a number of whom cultivate less than one acre, and as the available statistics do not include these small units, it is difficult to arrive at a reasonable estimate of the total acreage and production of these crops. One interesting feature of these comparatively new developments is the growth in the number of carriers and commission agents, who collect, arrange transport for, and sometimes bulk, the marketable produce.

Strawberries are the most important soft fruit in the area. The greater proportion of the 4000 odd acres is grown in the Isle around Wisbech. In the County the bulk of this fruit comes from small growers in the Cottenham, Willingham, and Histon districts. Varieties in favour are Royal Sovereign, Sir Joseph Paxton, Oberschelien, and Brenda Gautrey. Gooseberries still take second place in point of acreage, even though there has been a considerable falling off in the past few years. Black and red currants and raspberries were also widely grown at one time, but their cultivation is now confined to a few of the smallest holdings.

A feature of the last decade has been the development of bulb growing in the Wisbech district and of the culture of flowers, e.g. pyrethrums, scabious and outdoor chrysanthemums, in the Cottenham, Willingham, and Fordham areas. Here and there, too, nurseries have been established for the raising of fruit trees and shrubs.

A number of growers has recently erected glasshouses for the production of tomatoes, bulbs, forced mint, and indoor chrysanthemums, and though production has not yet reached large figures, development in this line is taking place steadily year by year.

Recently the Land Settlement Association has acquired two estates in Cambridgeshire and these have been equipped for the production of certain fruits and vegetables both indoors and under glass.

Over 10,000 acres are planted with top fruit, plums predominating in the County and apples in the Isle. On the whole, the climate is not all that might be desired, especially for apples, because the high winds and late frosts occasionally cause serious reductions in the crop yields. River's Early, Czar, Victoria, and Monarch are the more common varieties of plums; and of apples, most of which at present are of the cooking varieties, Bramley Seedling is a favourite, though Emneth Early was more popular in the old days. There is now a slow but steady change from culinary to dessert apples. An important area of fruit (particularly greengages) is to be found in the parishes of Melbourn and Meldreth, and other scattered areas of top fruit occur on the chalk soils in the south of the County, and on the gravels, peats, and skirt-soils near Burwell, Exning, and Fordham, apart from the larger areas around Cottenham, Histon, Rampton, Willingham and Wisbech.

Possibly in no other direction has research yielded such striking results or suggested such revolutionary changes as in fruit production. There is a definite tendency to grub up old orchards, planted before this new knowledge was available, and to replant with newer varieties on improved stocks and under conditions more conducive to the control of the many pests which do so much to limit the production of high-grade fruit. Many of the older orchards were mixed plantations of plums and apples, an unsuitable combination under modern methods of management.

## LIVE STOCK

The total number of *Horses* shows a steady decline from some 33,000 in 1913 to 20,000 in 1937 (Fig. 36), an experience which Cambridgeshire shares with the other arable counties. Though this decline has been more marked in the County than in the Isle, the replacement of the horse by the internal

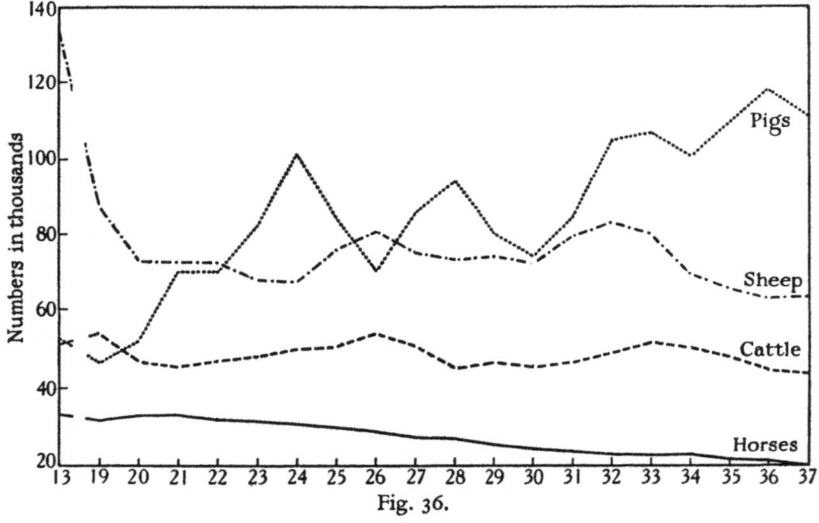


Fig. 36.

Main Live Stock, 1913–37 (Ministry of Agriculture Statistics).

combustion engine has been taking place quite as quickly in the latter part of Cambridgeshire. The horse still proves an essential supplement to the farm tractor, and now works even more constantly and effectively than formerly. In fact, the introduction of newer types of light-draught implements and of pneumatic tyres for farm carts and, in certain cases, the construction of concrete farm roads have had much the same effect as if there had been evolved by breeding and selection an animal of higher horse power.

No less apparent, than the decline in numbers, is the steady improvement in the type of horse to be found on the general farm. Pedigree breeders of Shires, Suffolks, and Percherons, the three most common breeds, find a good demand from local farmers as well as from those in other industries where a good heavy draught horse still proves invaluable. Quite a few, too, of even the smaller farmers, who have adequate facilities for rearing, breed one or more foals each year and assist in maintaining the high reputation gained during the last century by Cambridgeshire breeders of heavy horses.

The breeding and training of race horses is a feature on a considerable stretch of land around Newmarket both on the Cambridge and the Suffolk sides of the border. The greater part of this area, apart from Newmarket heath, is laid out in neat grass paddocks surrounded by shelter belts of trees, which give a distinctive appearance to a wide tract of land that otherwise would be featureless and rather bleak. Good paddock management, a matter requiring considerable skill and experience, is aided by the grazing of cattle in the summer and by the production of manure from yard-fed stock in the winter. Though localised, this industry is one of considerable importance to those who farm in the surrounding districts from the point of view of the demand for certain products of the farm and its requirements for labour.

The *Cattle* of Cambridgeshire are predominantly of the Shorthorn type, but dairy herds of Red Polls, British Friesians, and Jerseys, are to be found here and there, while Aberdeen Angus crosses occupy a number of the fattening yards and boxes. The total number of cattle has changed very little since 1913, though in recent years there has been some decline. Marked changes, however, have occurred in the cattle distribution and in the relative importance of the two main products—milk and beef. The number of cows in milk has increased slightly in the County and remained fairly constant in the Isle, but there is evidence that the quantity of milk coming on to the liquid market has increased very considerably, due mainly to three changes in practice, viz. the better management of the dairy herds; the almost complete suspension of farm butter-making; and a restriction of calf rearing and of the amount of milk fed to calves. To-day, Cambridgeshire dairy herds supply the local requirements for liquid milk, and they yield, in addition, an exportable surplus, much of which is consigned through local depots to London. These changes have naturally had their effect on the general farming system in certain areas, and there are many instances of the conversion of open yards, feeding-boxes and outhouses to cowsheds and milk and sterilising rooms.

Beef production in the same period has declined, partly through the utilisation of bullock-feeding yards for pork and bacon production and partly through the change over to milk production. There has been a steady decline in the number of older cattle fed for beef, especially in the Isle. The majority of the animals is now marketed at or under two and a half years of age and only a relatively small proportion at lighter weights or when fully mature.

The *Sheep* numbered, in 1913, some 108,000 in the County and 26,500 in the Isle. In 1919 the respective numbers were 74,500 and 14,000, and in 1937, 53,000 and 7500. These figures include a large number of lambs nearly fat and take no account of those stores which are purchased, fattened and sold during the winter season. Though there is little doubt that now more stores are purchased than formerly, the figures may be taken to indicate a general decline during the post-war period. This is an experience not uncommon to counties where formerly a high proportion of the sheep was maintained on arable land, and it represents a change in practice that may have some effect on the maintenance of fertility of the lighter soils.

On the chalk soils there are many parishes, which contained five or six large breeding flocks of folded sheep ten or twenty years ago, but which now can boast of only one or two flocks of much-reduced size. Important factors underlying this change undoubtedly have been the relatively high cost of labour and the desire, often the dictate of necessity, to grow a large acreage of direct cash crops. The introduction of sugar beet, a cash crop with a useful feeding residue in the form of tops, certainly has tended to check the decline in the arable sheep numbers during winter, for though ewe flocks may have been dispersed, it is not an uncommon practice to fold off the tops with store sheep.

The decline in the numbers of breeding sheep, however, cannot wholly be attributed to the reduction of the arable flocks. At least four factors have tended to a reduction of grass sheep, viz. (*a*) the more widely held opinion that, within certain limits, the lighter the sheep stock the better the health of the flock, (*b*) the increasing need for an adequate drainage of much of the heavier grassland, which must take precedence over its improvement by manuring and stocking, (*c*) the reduction in the number of grass orchards formerly grazed by sheep, (*d*) the replacement of sheep by pigs in many of the orchards which remain in grass.

Most of the arable flocks are of the Suffolk breed. In a number, kept pure, the breeding of rams for sale is an important item of the gross receipts from the flock; in others, the ewes are crossed with rams of another Down breed, usually the Hampshire. The Half Bred predominates on the grass farms. Generally the ewes of this breed are crossed with a Suffolk for the production of fat or store lamb. Scattered flocks of Hampshires are also maintained on arable land; South Downs run partly on grass and partly on arable land; while cross-bred ewes of various types are kept on a few of the grass farms.

The number of *Pigs* shows wide variations from year to year. Since 1932 the numbers have been considerably in excess of 100,000, which is double the figures of 1913 and 1919. No one factor is wholly responsible for this increase, but the greater part undoubtedly is due to the fact that it has been more profitable to stock the yards with pigs than with fattening cattle.

The stimulus to production given by the Pig Marketing Scheme has had effects on the changes in method of pig management. To-day, pigs are kept under very varied conditions, e.g. some are kept tethered out of doors all the year round, while others spend most of the year in a modern type of Danish piggery. Many barns, horse stables, and cattle yards have been converted for the use of pigs, and in all such alterations an important consideration has been to secure a layout which would allow of the maintenance of the largest number of stock per unit of labour.

Most of the store or fat stock coming on the market from Cambridgeshire farms are pure Large Whites or of the Large White-Large Black cross. There are numbers of pedigree breeders who favour these two breeds and who find a ready demand for the animals which they offer at their annual sales. There are also a few herds of the Essex and Middle White breeds.

*Poultry Keeping* has become an increasingly important branch of agriculture in Cambridgeshire as in other counties. It is mainly an activity of small-holders and general farmers, who derive only a part of their income from this source; but the number of specialist poultry farmers has increased considerably in recent years. Most of these keep flocks of from one to three or four thousand birds, and are interested both in egg production and table poultry. There are few larger units or special hatcheries.

## CONCLUSION

The post-war years have seen more far-reaching changes in Cambridgeshire agriculture than in any other short period of its history.[1] It is too early, yet, to see clearly the effects of these changes on the fertility of the soil or upon the livestock industry. There is little doubt that the scarcity and high cost of labour have led farmers to think more and more on the lines of mechanisation and to consider restricting the production of certain commodities which make heavy demands on labour. A short time ago it was not uncommon to find a number of farmers who were inclined to the

[1] See p. 135 above.

belief that soil fertility could be maintained, in successful arable farming, without the use of animal manure. To-day, this view receives little support. Though the labour problem is still as acute as ever, the tendency definitely is towards systems of balanced farming. Although some farmers may continue to deprecate the disappearance of the arable sheep flock and of the yard filled with fattening bullocks, it should be remembered that the larger numbers of folded store sheep and fattening pigs are proving a compensating factor in the maintenance of soil fertility.

Cambridgeshire as a whole may be regarded as a fertile county, but it includes considerable stretches yielding only a strictly limited amount of produce, chiefly because of badly drained soils and the lack of hard-bottomed roads. Field drainage, indeed, may be classed as one of the major problems of local agriculture. Blockages in small streams, overgrown ditches, bad outfalls (and, in consequence, blocked field drains) take a heavy toll on the crops over large tracts of land, particularly on the heavier soils. Most of the land in this condition has already been tile-drained within the last seventy or eighty years. A number of these systems may not be efficient, but there are many instances where the old drains function really well when given the opportunity. It is unfortunate that the greatest need for attention arises in those very districts where, through a variety of circumstances, the landlord or the tenant is not in a position to undertake the necessary work. The use of the mole drainer is helping to some extent, but this is not solving the sometimes greater problem of removing the water from the drained area.

There are also wide stretches of land (including some of the most fertile), and many farms, set two or three miles from a hard road and served only by a muddy track. Farmers so situated find it necessary to limit their production very largely to those commodities that can be carted off the farm during the drier months of the year. It is not unreasonable to suggest that over large areas the value of the crops might be increased by quite 50 per cent if field drainage were adequate and if roads permitted easy access all the year round.

## (C) REGIONAL TYPES OF FARMING

### By R. McG. Carslaw, M.A., PH.D.

Within the County boundaries there are marked contrasts both in the organisation and in the productivity of the farming. These contrasts are due primarily to differences in the nature of the surface soils, of which there is a remarkable variety.[1] Four major districts, comprising three-quarters of the land area of the County are to be discerned: (1) the chalks in the south and south-east, (2) the clays in the south-west, (3) and (4) the peats and silts in the middle and north (see Fig. 29). A brief comparative description of farm organisation in these districts, based on surveys during the 1931–36 period, will provide some indication of the principal types of farming in the County at this time.

(1) *The Chalk Soils.* In this area, farms and fields are large, many of the latter extending to more than 100 acres. The working capital here required[2] for stock, crops, and equipment (excluding value of land and buildings) approximates £12 per acre; gross income amounts to £10–£11 per acre; and employment is at the rate of three workers per 100 acres. Of the gross income approximately half is derived from crops. Rents average roughly 21*s.* per acre. Little more than 10 per cent of the farmed land is under permanent grass, and a common rotation for the arable land is (1) sugar beet, (2) barley, (3) barley, (4) seeds, (5) wheat. Barley is the principal cash crop, and excellent malting qualities are grown. Wheat and sugar beet are both important sources of income, while clovers and sainfoin are the principal short ley crops.

Sheep are the type of live stock traditionally associated with this district, and, formerly, large flocks of the heavier breeds (e.g. Suffolk) were kept for manuring and consolidating the arable fields. But the high labour costs entailed by close-folding, the decline in sheep prices, and the relatively more favourable returns offered by growing sugar beet in place of "sheep keep", have contributed towards reducing this practice. Indeed, on some farms sheep as an aid to soil fertility have been entirely superseded by artificial fertilisers and green manuring with rape or mustard. A relatively large number of pigs is kept to consume tail barley, and to convert straw into dung. Some cattle are yard fed during the winter months, going out fresh or fat in the spring, but in recent years low prices have kept many "yards" empty. Dairying is confined almost entirely to farms situated near villages, where an opportunity for retailing occurs.

Mechanisation in crop cultivation has here proceeded comparatively

[1] See Chapter ii.

[2] Cambridge University Farm Economics Branch, *Report* 24 (1937).

rapidly in recent years, partly, no doubt, owing to the presence of large fields, and partly also because of the extensive areas of cereals (three-fifths of the arable area) which are grown. Tractors and modern large-scale tractor equipment are now comparatively common.

(2) *The Heavy Clay Soils.* This is the least productive and most depressed agricultural district in the County. The soil is heavy and intractable; much land is in need of re-drainage; road facilities are in many places inadequate; derelict and semi-derelict fields are not uncommon.[1] In one parish, in 1932, it was found that out of nine holdings over 20 acres in size, five were uncultivated, one was vacated during the year as a result of bankruptcy, and two changed hands within the year owing to the financial difficulties of the occupiers. The Wheat Subsidy Act of 1932 gave a new lease of life to farmers in the district.

A survey carried out during 1933 showed that roughly 60 per cent of the farmed land is arable.[2] Working capital in stock, crops, and equipment averages some £7 to £9 per acre, gross incomes £5 to £7, rents 15*s*., and employment little more than two workers per 100 acres. The soil is so heavy that opportunities for diversified cropping are limited, and high-value crops such as sugar beet and potatoes can seldom be grown. The barley produced is generally of poor quality. There is a comparatively high proportion of bare fallow, and a considerable amount of cross-cropping. A not uncommon rotation is (1) wheat, (2) seeds with a bastard fallow, (3) wheat, (4) bare fallow. Beans are the chief fallow crop, and a limited area of field peas is also grown. Red clover is the main "short leyer" crop, and second cuts are frequently taken for seed. Trefoil and sainfoin are also found amongst the leyers. Wheat is the principal cash crop, but where soil conditions permit small areas of fruit, potatoes, and market-garden crops are grown.

Fewer live stock, particularly smaller numbers of pigs and sheep, are produced here than on the chalk soils. The grazing season is relatively short, perhaps owing to poor drainage and management, and the land "poaches" badly in winter. In recent years, some movement has been made towards developing a system of cropping involving long leyers of 3–5 years in place of the usual 1-year leyer. This lengthening of the rotation into, say, 4 years plough followed by 4 years grass, appears to hold opportunities for improving the organisation of farms in the district. But there is an acute shortage of working capital amongst the farmers, and improvements or adjustments necessitating capital outlay, such as fencing and provision of water, can only be slowly adopted.

[1] See pp. 56 and 131 above.

[2] Cambridge University Farm Economics Branch, *Report* 22 (1935).

(3) *The Black Peat Soils.* The farms here tend to be rather smaller in acreage than those on the neighbouring uplands, but the organisation of production is much more intensive. Capitalisation, employment, and output per 100 acres are high. A survey[1] made in 1936 showed that an ingoing tenant requires some £25 working capital per acre of farmed land (arable plus pasture), that employment is at the rate of about five workers per 100 acres, and that gross sales amount to £20 per acre per annum. Rents, including the tenant's share of drainage rates, average 50*s.* per acre, while the labour bill approximates 90*s.* per acre. These various measures are roughly double the comparable data for upland farms in the south of the County. Road facilities are in many cases poor, surfaces being unmetalled and virtually impassable in winter, but the numerous waterways provide an alternative means of transport.

The usual rotation is (1) wheat, (2) potatoes, (3) sugar beet. A few acres of oats for horse feed are grown, while other crops commonly found are celery, carrots, and mustard seed. There is very little temporary grass, and permanent pasture consists mainly of off-lying "wash" grazings, frequently flooded during winter. The outstanding cropping characteristics are the large proportion of the farmed area which is cultivated, the large proportion of the cultivated area which is devoted to cash crops, and the concentration on crops giving a high money output per acre, e.g. potatoes (see Fig. 34), sugar beet (see Fig. 35), and celery. Crop yields per acre are high, probably averaging one-third above that on upland farms; yields of 15 tons of sugar beet, 12 tons of potatoes, and 50 bushels of wheat, per acre are not uncommon. These good crop yields are no doubt mainly due to the inherent fertility of the soil, but liberal applications of artificial fertilisers and good management contribute to the result. Most of the dung and fertilisers is applied to the sugar beet, potatoes, and celery. Potatoes, for example, are commonly dressed with 15–20 loads per acre of farmyard manure, plus 6–10 cwt. of fertilisers, while sugar beet may get 4–6 cwt. of fertilisers.

Of the gross sales from the black peat farms surveyed in 1936, crops accounted for three-quarters and live stock for one-quarter. Both the absolute and relative importance of crops on the peats is thus very much greater than on the chalks and clays in the south of the County. Sugar beet and potatoes are the two major items of revenue, and together they amount to more than half the total receipts, with wheat coming third in importance. In cash values, pigs are the most important livestock enterprise, with cattle second.

[1] R. McG. Carslaw, "Farm Organisation on the Black Fens of the Isle of Ely", *Jour. Roy. Agric. Soc.* xcviii, 35 (1937).

A common practice in the cattle management is to buy stores in the early spring, yard feed them for a few weeks before sending them to summer grass on a "wash" pasture, and to bring them back to the yards in October for fattening off during the winter. Where no cattle are fattened pigs are usually kept to tread down the straw. Sheep are conspicuous by their absence, and very little milk or poultry production is undertaken. Although horse-breeding is associated with this district, sales of horses amounted to only a little over 1 per cent of gross incomes on the farms covered by the survey.

Considering the types of crops grown, the farmers are remarkably independent of imported casual labour, as the wives and families of the regular employees commonly assist with seasonal operations. Beet thinning, potato and celery planting, and the beet and potato harvests are generally let out at piece rates, and individual families of workers may earn substantial sums at certain times.

Fenland farming depends, of course, on a complex system of artificial drainage. On many farms, however, the drainage appears to be satisfactory, the most usual complaint being that the water level is kept too low during the summer months. Surface water-logging seems to be a more serious difficulty to the farmers than any defect in the main drainage system. Particularly on land where, owing to "wastage" of the peat,[1] the underlying clay is now close to the surface, pools of water form after heavy rain. The crop will quickly deteriorate in these patches unless the surface water is removed, and this is generally done by ploughing or hand-digging water furrows to the nearest ditch.

The practice of "claying" the peat soils[2] is less frequently practised than in the past owing to the high labour costs involved. In some cases wastage, however, has proceeded so far that the clay is now being ploughed up and mixed with the peat during the ordinary field cultivations.

(4) *The Silt Soils.* These extraordinarily fertile alluvial deposits vary from a light to a heavy consistency according to the percentage of clay. The economic organisation of farms in the district is in many ways similar to that on the peats, but production is even more intensive; and capitalisation, output, and employment are generally higher. Rentals range from £3 to £5 per acre. Compared with the black peats, less sugar beet is grown, and potatoes (chiefly Majestics and King Edward's) are a relatively more important crop; further, the quality of the potatoes, and therefore

[1] See p. 186 below.

[2] See p. 120 above. It is generally carried out by digging narrow ditches across the field down to the underlying clay, throwing the clay out and spreading it evenly over the field, and then filling up the excavations to make all reasonably level. In 1936 the operation cost from £10 to £15 per acre.

the price per ton received by growers, is better. The rotation approximates to (1) potatoes, (2) sugar beet, mustard for seed, and various root seeds, (3) wheat, but is widened by inserting oats, peas, or clover where desirable. Considerable areas of root seed crops (turnips, swedes, mangolds and sugar beet) are also grown. Fruit is important, particularly in the vicinity of Wisbech.[1] In 1936, over 3000 acres of strawberries, 3000 acres of apples, 1400 acres of gooseberries, and 1400 acres of plums were grown in the Isle of Ely, principally on these silt soils. Although some 500 acres of bulbs were grown in 1936 this industry, together with market-garden and glasshouse production, is less fully developed than in the Holland Division of Lincolnshire which lies immediately to the north of Cambridgeshire. Indeed, the silts which lie within Cambridgeshire are only a small part of the large compact silt area surrounding the Wash, and which includes the whole of the Holland Division.

(5) *Other Districts*. The four main areas already described cover approximately three-quarters of Cambridgeshire. The remainder of the County includes a number of small areas of varying soil types.[2] In particular, no account of the agricultural regions of the County would be complete without reference to the fruit-growing area immediately to the north of the town of Cambridge. This includes the parishes of Milton, Waterbeach, Landbeach, Impington, Histon, Cottenham, Rampton, Long Stanton, Willingham and Over.[3] Since the middle of the nineteenth century, a strong concentration of fruit growing (especially plums, apples and strawberries) has been developed here.[4] In more recent years, fruit has been supplemented by the introduction (often by underplanting the top fruit) of market-garden produce (asparagus, cauliflowers, broccoli, dwarf beans, and peas), and of cutting flowers (pyrethrums, scabious, iris, gladioli, asters, marguerites, gypsophila, etc.). Small-holdings of 20 acres or less, producing these intensive crops, are numerous in the district, while there is a large number of "part-time" holdings, of an acre or so, in the occupation of agricultural labourers and other wage-earners. Poultry and pigs are the most usual types of live stock, and are kept largely to produce manure and to utilise by-products.

[1] See C. Wright and J. F. Ward, *A Survey of the Soils and Fruit of the Wisbech Area* (1929). See p. 143 above.
[2] See Chapter ii, and Fig. 29.
[3] See J. F. Ward, *West Cambridgeshire Fruit-Growing Area* (1933). See p. 143 above.
[4] See p. 131 above.

CHAPTER ELEVEN

# THE INDUSTRIES OF CAMBRIDGESHIRE

## By F. M. Page, M.A., PH.D.[1]

DANIEL DEFOE, WRITING IN 1724–26, SUMMED UP HIS impression of Cambridgeshire by saying that "this county has no manufacture at all"[2] Although this is less true now than then, it is still not surprising that industry should occupy a subsidiary place in one of the most agricultural of counties. Chronologically, the industries of the County fall into three groups. In the first place, there are the extinct industries: thus during the seventeenth and eighteenth centuries the production of saffron was flourishing in the south-east of the County, there were two saltpetre factories at Cambridge and Barnwell, and bell-founding was carried on in Cambridge. During the nineteenth century came the digging of coprolites.[3] At this time, too, there was the activity of the various industrial schools; spinning establishments existed at Fowlmere, Soham, and Histon, and in this last village, stockings were also made; while at Wisbech and Linton hemp was made into rope. Perhaps the most interesting of these extinct industries was the manufacture of woad at Parson's Drove, some 6 miles from Wisbech. Working ceased in 1914, but the mill still survives. This mill together with two others near Boston are the last representatives in Europe of an industry that dates from pre-Christian times.

Secondly, come the industries with a long continuous history through the Middle Ages to the present day. The best examples, perhaps, are basket-making, printing and book-binding, quarrying for stone and clay. Finally, there are the industries of recent growth, conspicuous elements in shaping the modern economy of the County. In this class come the manufacture of sugar, the canning and preserving of fruit and vegetables, and the construction of scientific instruments and apparatus.

The account that follows does not profess to cover every commercial undertaking in the County. It can deal only with the most important and the most characteristic.

[1] For help in the preparation of this chapter, I am indebted to Mr F. J. Corbett, the Secretary of the Cambridge Chamber of Commerce, to Mr R. S. Whipple, past President of the Chamber, to Mr John Saltmarsh of King's College, and to the numerous firms who have given me information. I am indebted to the Editor (Mr L. F. Salzman) for allowing me to use material prepared for the *Victoria County History of Cambridgeshire*.

[2] D. Defoe, *Tour through England and Wales* (1724–26), Letter I.

[3] See p. 126 above.

*Agricultural Industries.* The basket-making and wicker industry of the County is very long-standing. One of the first things that struck Camden, in 1587, was the "willows in great abundance, either growing wild or set on the banks of rivers to prevent overflowing. It is of these that baskets are made."[1] This fen occupation survived the draining, and, to-day, the chief centres are Ely, Soham, Chatteris, Over, and Somersham (Hunts). Here are made wicker-chairs, bottle-containers, and every form of wicker-work. A different kind of basket is made at the Wisbech saw-mills. A good local market was provided by the surrounding fruit district. The Wisbech and District Fruit Growers Association bought up all the baskets, before the output increased sufficiently to supply a national market. The most important firm was Messrs Dewsbury Bros. Now, the British Basket and Besto Co. Ltd. carries on the industry, and the timber-working firms themselves have established departments for it.

Timber-working at Wisbech was established during the nineteenth century. The first cargo of foreign timber arrived at the port in 1824, brought by an English barque. The next hundred years saw considerable expansion. The leading firm in the development is Messrs English Bros, Ltd., and its raw material is mainly Norwegian. There are branches at Sutton Bridge, Boston, and Peterborough, as well as at Cambridge itself; and the firm has been a pioneer in the use of creosote oil for making wood weatherproof. The oil is forced into the timber by steam pressure in air-tight cylinders, and, impregnated in this way, it withstands the effects of damp without being coated either with tar or paint. Telegraph poles, gateposts, and railway sleepers, submitted to this treatment, have remained unrotted after fifty years' exposure.

Characteristic occupations that have lingered into recent times are the preparation of reed and sedge for thatching, the digging of turf for fuel, the making of hurdles from willows; Reach and Burwell were the main centres near Cambridge; but now these activities have almost disappeared. Another product of the soil provided material for the straw-plaiting industry. This once flourished in the south-west of the County at Little Gransden and Littlington, but now the only firms are at Cambridge, Ely, and Little Shelford. A more famous industry born of the soil has been brewing. Cambridgeshire barley was at hand, and it seems as if malt, beer, and ale were among the commodities for which the County was most famous throughout the middle ages into modern times.

*Sugar manufacture* is a more recent, but a more important industry. The encouragement of the sugar-beet industry by government aid in post-war years had important consequences for the County. The acreage under

[1] W. Camden, *Britannia* (1637 edition), p. 491.

sugar beet rapidly grew,[1] and, now, the Isle of Ely and the County of Cambridge together account for 66 per cent of the total acreage under beet in Britain. In 1924, a factory at Ely was built in the centre of this agricultural activity. In the beet "campaign" of 1933, Ely out of eighteen competitors came second (with Peterborough) in "rated beet capacity",[2] its figure being 240,000 tons; the factory came third in "through-put of beet", producing 272,264 tons.

In addition to the production of sugar, there are useful by-product industries. The beet tops are used for manure or for cattle feed; molasses produced during refining are sold for distillation or for fodder; beet pulp (fibre after extraction of the juice) also forms good cattle food, equivalent to eight times its weight in mangolds; finally, the lime sludge is used for manures.

*Preserving and Canning.* Fruit growing had long been famous in Cambridgeshire, but the danger of over-production was great. Without some method of preserving on a large scale, fruit that could not find a local market had to be left to rot as it stood. Mr Stephen Chivers and his sons, about the year 1873, decided to experiment with the surplus fruit of the small farm at Histon that had been held by their family since the beginning of the century. The first boiling took place in a barn that can still be seen. A Cambridge grocer, greatly daring, volunteered to dispose of the jam, and was apparently much surprised to find that it sold. Accordingly, in 1875, a small factory was built conveniently near the railway, in case the venture might justify distribution to wider markets. Improvements in equipment were steadily made. The Galloway boiler was introduced about 1885; and the introduction of electric light enabled fruit to be made into jam as soon as it was picked. At this time 150 workmen were employed. To-day, the Orchard Factory has between 2000 and 3000 employees; its estates cover 8000 acres; its market is world-wide. It is estimated that 100 tons of jam can be produced daily. To this initial manufacture, other commodities have been added—jelly tablets, custard and blanc-mange powders, mincemeat, and marmalade. Thus an even pressure of employment is kept up throughout the year, in and out of the English fruit season.

Messrs Chivers & Sons were also among the pioneers of the canning industry in this country. The first bottle of preserved fruit was produced in 1890, and the first "tin" of fruit in 1893. By 1931, a new factory was opened at Huntingdon to take over the canning of vegetables, the fruit being still treated at Histon. It is interesting to note that all containers, jam-pot covers, and boxes, etc. are made on the spot.

[1] See Fig. 33.
[2] *Report on the U.K. Sugar Industry* (Blue Book, 1935), Table xvi, p. 30.

Recently, a branch of Messrs S. W. Smedley & Co. has been established at Wisbech, the northern fruit and vegetable centre of the County. The main attraction was the strawberries of the Wisbech district, and the plums and greengages growing around Ely.

*Agricultural Implements.* In this agricultural setting, it is not unnatural to find a number of firms manufacturing agricultural implements. As early as 1884, the Falcon Works (John Baker Ltd.) at Wisbech were notable for the invention and manufacture of corn- and seed-dressing machines, at a time when mechanism was only slowly being introduced into agriculture. Firms like Messrs Kidd of Willingham and Messrs Lack & Sons Ltd. of Cottenham have also a long history as agricultural engineers; while the chaff cutters of John Maynard of Whittlesford have reached many parts of the world. Prominent, also, are Messrs Edwards & Sons of Wisbech and Messrs Macintosh & Sons Ltd. at Cambridge. The only iron foundry now in the district is that of John Hart at Cottenham.

*Quarrying.* Three geological formations are of importance in Cambridgeshire industry. The Chalk contributes flints, chalk for limeburning and for cement, and the soft building stone known as clunch. Surviving bursar's accounts show that when the Cambridge colleges were being built, considerable amounts of stone were obtained from Haslingfield, Barrington, Cherry Hinton, Reach, and Burwell. The stone for the Great Gate at Trinity came from Burwell and Cherry Hinton; that of the Gate of Honour at Caius came from Reach. Clunch was also much used in interior decoration; examples may be seen in the fan-tracery of the Lady Chapel at Ely Cathedral, or in St John's College Chapel. Quarries are still to be found along the line of this outcrop (Burwell Rock, or Totternhoe Stone) in the Lower Chalk. Many of the old quarries at Isleham and elsewhere are now disused, but clunch is still dug for road-making.

The Lower Greensand furnishes an easily dressed stone, known as Carstone, which has been used for houses and churches, but there is no great quantity, and it has seldom been carried for long distances.

*Brick and Cement Works.* The clays of Cambridgeshire have given rise to pottery and brickworks. The potteries have disappeared but the brickworks are very active. The bricks are of two kinds: the Gault produces a yellowish grey brick, very common in Cambridge itself; while the Jurassic Clays yield red bricks. At Cambridge, Ely, and Whittlesea, there are a number of well-established brickworks. Clay mixed with chalk also provides material for cement works at Shepreth, Meldreth, and Barrington, and for the British Portland Cement works at Coldham's Lane, Cambridge, with a weekly output of about 2000 tons. Then, in addition, there are several concrete manufacturers such as the Cambridge Concrete Co. of

Milton, with its own pit of gravel and sand, specialising in roofing tiles, blocks and bricks; the Cambridge Artificial Stone Co., dealing mainly with architectural specialities; the Atlas Stone Co., producing paving slabs and kerbing; and Messrs Tidnams Ltd. of Wisbech, concerned with a variety of concrete products.

*Printing.* Amidst much that is obscure and controversial, two facts stand out clearly in the early history of Cambridge printing: first, that John Siberch, a friend of Erasmus, started printing in 1521; and, secondly, that the University received clear authority to "print all manner of books" under the charter granted by Henry VIII in 1534. During the sixteenth and seventeenth centuries the primary policy of the University was to protect their printers' privileges rather than to develop the business of book distribution, and it was the common practice for Cambridge books to be sold through London booksellers. At the end of the seventeenth century, the University Press was organised as a University department. Large-scale reorganisation was undertaken by Richard Bentley, who secured the appointment of the first Press Syndicate; from 1698 to the present day, the Press has been governed by a body of resident graduates known as the Syndics of the Press. During the eighteenth century, the Syndics felt their way towards publishing as well as printing. Their chief stock-in-trade at this time consisted of Bibles and Prayer Books,[1] but some notable books, such as Newton's *Principia* and Browne's *Christian Morals*, were also published in the early part of the century. Stereotyping was introduced about 1734 and an improved method early in the next century. The earliest printers carried on their work in various parts of the town, and the first University printing house was on the site of the present lodge of St Catharine's College. In 1804 a new building was erected on the south side of Silver Street, and in the course of time the Press has gradually absorbed the whole of the site between Silver Street and Mill Lane, the most prominent feature being the Pitt Press, erected in memory of Pitt in 1833, and recently reconstructed. A publishing department was inaugurated in London in 1873, and the University Press now employs about 320 men in the printing house at Cambridge and about 120 in Bentley House, the headquarters of its London publishing. Its catalogue contains the titles of about 5000 books and journals which are distributed to booksellers throughout the world from Bentley House. All these are issued with the *imprimatur* of the Syndics of the Press.

[1] In common with the King's Printers and the Oxford University Press, the Syndics retain the privilege of printing the Authorised Version and the Book of Common Prayer—a privilege exercised by virtue of the charter granted by Henry VIII in 1534, and confirmed by Charles I in 1628.

Apart from the history of the University Press, there is very little authentic record of printing in Cambridge[1] until near the middle of the eighteenth century when a weekly newspaper, *The Cambridge Journal & Weekly Flying Post*, was published in September 1744. Eighteen years later, in 1762, *The Cambridge Chronicle* was first issued, and about four years afterwards *The Journal* was incorporated. This paper had no rival until 1839, when *The Cambridge Advertiser* (which subsequently became *The Independent Press*) first saw the light. These local newspapers were mainly responsible for general commercial printing, although in the early years of the reign of Queen Victoria, one or two small printers established themselves in the town, but their activities never assumed large proportions. *The Cambridge Express* also came into being; but with the advent of *The Cambridge Weekly News* in 1887, the three other local newspapers were absorbed, *The Chronicle* being the last to be incorporated a few years ago. The printing department of the latter now survives as "St Tibb's Press". A few of the old private firms remain without having shown much expansion, with the exception of Messrs W. Heffer & Sons Ltd., who started by taking over the small jobbing section of *The Independent Press*, and who now have one of the most up-to-date works in the Eastern Counties.

*Instrument-Making.* When Sir Michael Foster was appointed to the University Chair of Physiology in 1883, he found a startling lack of medical equipment of British and modern design; most instruments needed to keep pace with medical discovery had to be imported from German firms. Consequently he started to design and manufacture instruments on a small scale with the aid of two former pupils, Dr Dew Smith and Mr Francis Balfour. Soon, the co-operation of Sir Horace Darwin was obtained, and this was the beginning of the Cambridge Instrument Company Ltd. It was not until 1895, however, after the retirement of the senior partner, that the business registered itself as a company under the chairmanship of Darwin. Among the important inventions of those early days were the bifilar pendulum form of seismograph, and the rocking microtome for the rapid preparation of specimens for the microscope; then again there was the thread-recorder for marking the path of a moving pointer.

After the changes in reorganisation, the business was removed from St Tibb's Row to Chesterton Road, where it has remained, adding block to block, until the present day. Darwin took a leading part in the study of aviation. In 1912, he was appointed a member of the Advisory

[1] There are of course printers at work outside the town of Cambridge itself. Thus "the earliest newspaper bearing a Wisbech title—the *Lynn and Wisbech Packet*—came into existence on January 7th, 1800". For the subsequent newspaper history of the town, see F. J. Gardiner, *History of Wisbech and Neighbourhood* (1898), pp. 65–74.

Committee on Aeronautics, and the Company began to produce height-finders and instruments to locate the presence of aircraft. When war broke out, experimental effort was redoubled. A special thermometer was produced for testing the temperature of water in an aeroplane radiator—a very important invention when flying at great heights became normal. Other aircraft instruments were also produced.

Another Cambridge firm specialising in the manufacture of scientific apparatus was the "Granta" Works founded by W. G. Pye about 1897. In addition to supplying equipment to laboratories for teaching purposes, many pieces of apparatus were manufactured for some specific experiment, and the demand grew. Graduates from Cambridge, in equipping laboratories elsewhere, looked for apparatus similar to that which they had used in their training. Particular attention had been paid to electrical instruments, and in order to provide work for men returning from active service in 1919 and 1920, apparatus was developed for teaching the principles of wireless telegraphy. Broadcasting commenced, and, soon, the teaching panels were in great demand for listening-in. With their circuit lines engraved in white on ebonite panels, they took principal place in many drawing rooms. Later developments, and especially the advent of the portable receiver, resulted in very great extension of wireless production, and in 1929 it was decided to separate the two activities. The Radio department was disposed of to the Pye Radio Co. Ltd. The original business of scientific instrument-making was carried on in modern premises in Newmarket Road; later, an Aeronautical Instruments Section was added, retaining the style as W. G. Pye & Co. Ltd.

A more recent instrument-making firm is Messrs Unicam Ltd. This was started in 1933 at St Andrews Hill, but has recently moved to enlarged premises at Arbury Road. Finally, Clifton Instruments Ltd., founded at Bristol in 1929, was transferred to Cambridge in 1938. This is concerned with physiological instruments.

*Paper-Making, etc.* The Sawston Paper Mill is one of the oldest paper mills in the country, and the only one now in existence in East Anglia. The mill is known to have been making paper in 1664, and possibly the manufacture had been carried on from a much earlier date.

The name of Fourdrinier is found in association with the Mill as early as 1780, and it is certain that one of the earliest paper-making machines in the country was installed at Sawston. In 1836 the Mill passed into the Towgood family, who made paper continuously from that date to 1917, and who built up a reputation for a very high grade of paper. In 1917, the Mill was incorporated as a Limited Company, under the name of Edward Towgood & Sons Ltd., and passed into the possession of the well-known London Stationers, Spicer Bros. Ltd., now Spicers Ltd. Spicers Ltd. have developed on this estate a flourishing group of factories, where Woodpulp Containers, Envelopes, Waxed Wrappings, D'oyleys, Account Books, and other products of the stationers' craft are made.

Forming part of the factory extension at the Sawston Mills is the activity of Dufay-Chromex Ltd. Over the last ten or twelve years, experimental work in connection with the manufacture of non-inflammable colour film, under the Dufaycolor process, has been developed, and the manufacture of this well-known colour film is proceeding.

There are also other activities at Sawston. The manufacture of chamois ("shammy") leather has been carried on for over a hundred years.[1] Like that of paper, it was no doubt started here because of the good supply of water and the easy means of transport. The refuse from the skins goes to make soap, glue, dubbin, or manure for fruit trees. Glove-making is also carried on, making Sawston a unique example of an industrial village in Cambridgeshire.

*Miscellaneous Industries* have sprung up spasmodically in Cambridgeshire with no particular reason for their location. Wisbech provides example of an extensive tent-manufacture, which was of great importance in providing equipment for the Boer War, as well as for flower-shows, fairs, garden-parties, and camps. In this town there is also a label factory (Messrs Burall) which was one of the first firms to produce the clip-on type of label as opposed to the more usual eylet-and-string model.

At Littleport, Messrs Hope Bros. have recently set up a factory for shirt-making giving employment to over 300 people; at Whittlesford there is vinegar-brewing; while at Whittlesford and Pampisford there are artificial manure works.

In Cambridge itself there are some famous firms manufacturing brushes—the Cambridge Brush Company, the Kleen-e-ze Company, and the Premier Company. The Cambridge Tapestry Company is important for the special study that has been made of the repair of ancient fabrics and upholstery, by means of which many medieval treasures have been saved for posterity. Finally there must be mentioned the firms of box-manufacturers, turners, and furnishers, the Cambridge Metal Stamping Company, and the Cambridge Gas Company. But this does not exhaust the list.

[1] For a description of the process, see T. M. Hughes and C. Hughes, *Cambridgeshire* (1909), p. 103.

CHAPTER TWELVE

# THE GROWTH OF CAMBRIDGE

By J. B. Mitchell, M.A.

TWO ESSENTIAL ELEMENTS NEED STRESS IN THE SITUATION of Cambridge: its position on the Cam at the junction of Fenland and Upland and its relation to the open chalk country and gravel terraces controlling the land routes. The Cam, navigable from Lynn to Cambridge, was a main artery of communication through the Fenland: sea-going vessels were still discharging their goods at Cambridge quays and hithes in 1295, while river traffic remained of great importance until the competition of the railways ruined the watermen in the nineteenth century. Cambridge also owes much to its land routes. A slight transverse fold in the south-easterly dipping chalk throws a finger from the chalk escarpment north-west across the valley of the Cam,[1] the severed tip of this finger forming the chalk outlier of Castle Hill. This ridge capped and extended by gravel-spreads provided a ford across the river, and constituted a south-east—north-west land route from East Anglia to the Midlands crossing the north-east—south-west river route at Cambridge. The main Roman road of the area, the *Via Devana*, exactly followed this ridge: the modern roads south of the river approach along the gravel terraces of the valley but continue north-west along the ridge to-day (Fig. 37).[2]

The site itself, where chalk and gravel approach the river, afforded the essentials of solid banks for bridging and dry ground for building, and was partially protected by the sweep of the river and its marshes. The distribution of the gravels, which largely determine the minor elevations (compare Figs. 37 and 38),[3] assume a great importance on a site as low and liable to flood as Cambridge. The gravels within the meander consist essentially of the Higher Terrace gravels of Trumpington to the west and of Barnwell to the east rising to 50 ft., separated by a spread of Lower and Intermediate Terrace gravels lying approximately at 30 ft. O.D. Along the centre of the

[1] I.e. the Gogmagog Hills which are shown up clearly in Fig. 8.

[2] On Fig. 37, the 100 ft., 50 ft. and 20 ft. contours have been traced from the Ordnance Survey 6 in. sheets: form lines, at 10 ft. intervals, have been interpolated from the O.S. 25 in. plans (1925 edition). The unequal intervals of the layer colouring have been deliberately chosen for comparison with the built-up areas.

[3] Fig. 38 is based on the 1 in. Geological Survey sheets (Drift edition) of the area. For access to the 6 in. map of the southern part of the area, and to a map of the Geology of Cambridge, by A. J. Jukes-Brown, I am indebted to the Director of the Geological Survey.

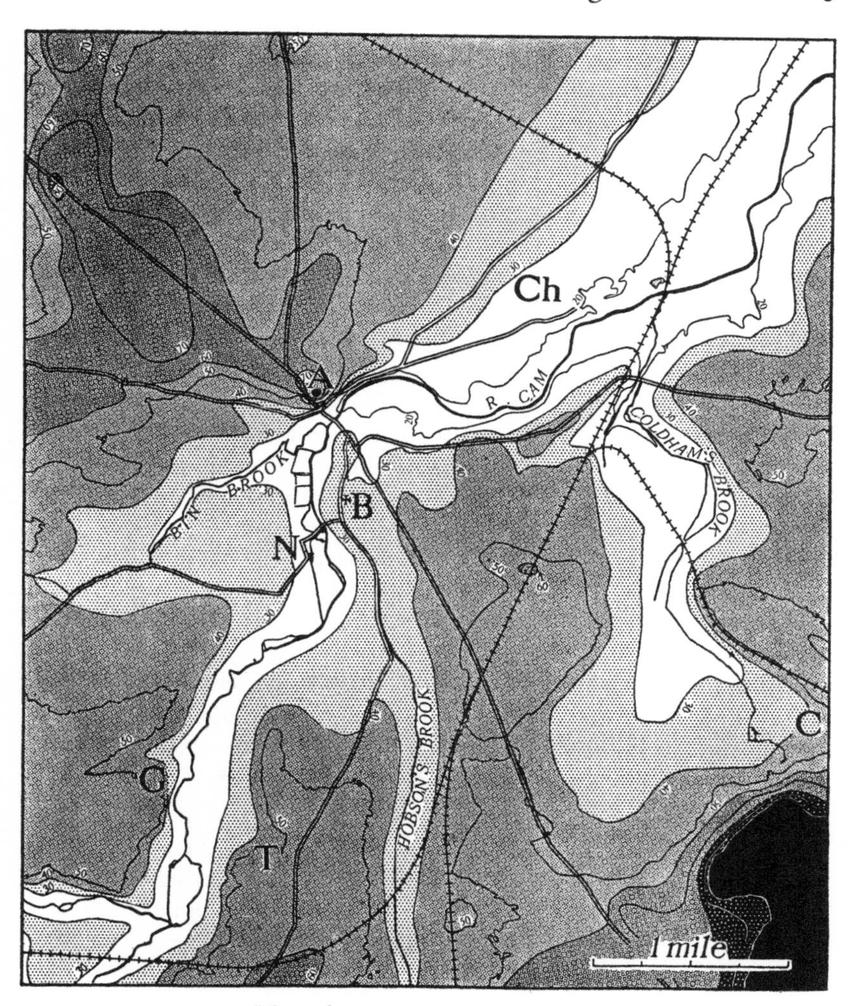


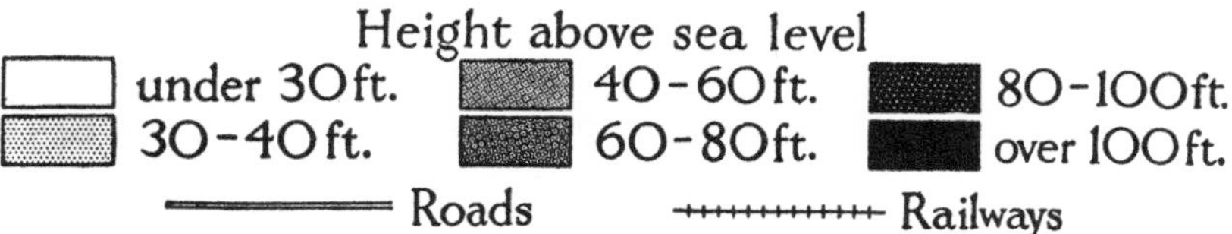


Fig. 37.

The Cambridge Area: Relief.

A = The Castle. B = St Bene't's Church. N = Newnham. G = Grantchester. T = Trumpington. C = Cherryhinton. Ch = Chesterton. For sources see footnote 2, p. 162.

Intermediate Terrace gravels runs a depression occupied by Hobson's Brook in the south and followed in part by the King's Ditch in the north. To the east, these gravels are separated from the Higher Terrace gravels of Barnwell by an outcrop of Chalk and Gault which coincides markedly with the eastern belt of open land formed by the University Sports Ground (Fenner's), Parker's Piece, Christ's Pieces and Butt's Green. North and west of the river are gravel-spreads equally important to the settlement of the area; the Higher Terrace gravels at Grantchester are comparable to those of Trumpington; Intermediate and Lower Terrace gravels stretch from thence to the valley of the Bin Brook and are separated by an outcrop of gault from the high-lying Observatory gravels of Castle Hill. East of Castle Hill lies a wide spread of Intermediate and Lower Terrace gravels reaching to Chesterton and beyond. The Cam River is bordered by alluvium, once marsh, but now largely drained and raised to form a belt of open land to the west and north of the town, comprising Sheep's Green, Coe Fen, the Backs, Jesus Green, Midsummer Common, Chesterton Fen and Stourbridge Common.

The gravels not only afforded well-drained building sites, but, paradoxically, gave the early town an ample, if not always sanitary, water supply. The underground seepage of water towards the Cam, held up by the impervious Gault, was tapped by shallow wells in the gravels, and provided until the seventeenth century a water supply considered adequate for all needs. In modern times,[1] two sources of water supply have replaced the easily contaminated surface wells; artesian water from the Lower Greensand and, more important, water from the Chalk.

## THE MEDIEVAL PERIOD[2]

Evidence of a pre-Roman settlement at Cambridge is lacking, but Bronze Age finds, beakers, and burials, clustered near the ford, point to its use at this date; and in Roman times a fortified settlement mounted guard on Castle Hill over the important crossing at its foot. In Anglo-Saxon times, a flourishing settlement, or more probably settlements, developed in association with the river bend: (1) to the south of the river the Saxon

[1] (*a*) "This year [1610] the Town and University completed a new river from a place called Nine Wells in the Parish of Great Shelford to the Town of Cambridge." C. H. Cooper, *Annals of Cambridge*, iii, 36 (1845). "This year [1614] Henry King and Nathaniel Craddock with the King's sanction, and at the joint charge of the University and the Town, undertook to convey water by pipes from the new river to the Market Place, and there to erect a conduit of stone." *Ibid.* iii, 62.

(*b*) The Cambridge Town and University Waterworks Co. was formed in 1853.

[2] The theories regarding the early history of the town are discussed in an admirable paper by H. M. Cam, "The Origin of the Borough of Cambridge", *Proc. Camb. Antiq. Soc.* xxxv, 33 (1935).

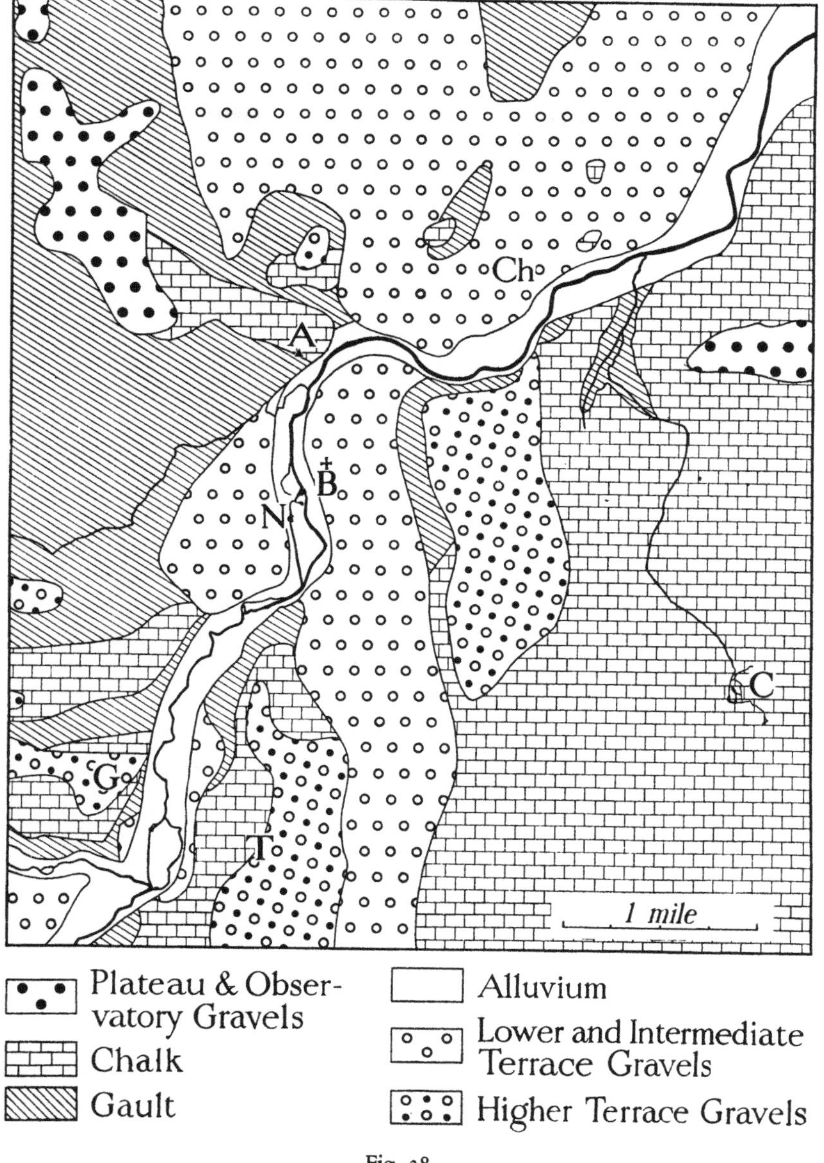


Fig. 38.

The Cambridge Area: Drift Geology.

A = The Castle. B = St Bene't's Church. N = Newnham. G = Grantchester. T = Trumpington. C = Cherryhinton. Ch = Chesterton. For sources see footnote 3, p. 162.

tower of the Church of St Benedict bears witness to the early date of the occupation of the gravels of Market Hill and Peas Hill, between the Cam and the King's Ditch depression; (2) to the north of the river, Castle Hill had also been early built upon; Domesday Book records that there were fifty-four tenements in Castle Ward. There also appear to have been two smaller settlements; (3) to the west on the rising ground of the river terrace at Newnham, the Mill at Newnham is mentioned in the Domesday Survey; and (4) to the east on the northern edge of the gravels at Barnwell. The main ford of the river at the foot of Castle Hill, where gravel and chalk afford firm banks, was early bridged; the Great Bridge of the documents was situated here. The Small Bridges, near the Mill Pool, connecting the settlement within the meander with that of Newnham, are also of early date. The medieval town was formed by the expansion of the two centres at Castle Hill and at Market Hill; but it was not until modern times that the settlements at Newnham and Barnwell were completely absorbed.

The medieval town so formed may have been bounded eastwards by the King's Ditch (see Fig. 39), cut, most probably, primarily for the defence of the crossing, not for the safety of the settlements. By the thirteenth century, the town had grown beyond these limits: the parish of St Mary the Less to the south and that of St Andrew the Great to the east both lay almost entirely outside the town as defined by the King's Ditch, and both had a considerable population. The further extension of medieval Cambridge was, however, confined on the one hand by the alluvial marshes of the river, and, on the other, by the inviolability of the town fields. Already by the end of the thirteenth century, the edge of the gravels was being raised and drained to provide extra building sites without sacrificing valuable agricultural and meadow land: the chapel and infirmary of the Hospital of St John (later the site of St John's College) and the nunnery of St Radegund (later the site of Jesus College) encroached on the alluvial land of the western and northern slopes of the Intermediate gravels.

The University, already powerful at this period, did not, however, possess elaborate buildings. The great period of University and Collegiate building belongs to the fourteenth and fifteenth centuries. By this date, most of the desirable gravel sites within the King's Ditch boundary had been occupied, and thus the Collegiate buildings fall into two groups: those upon good gravel sites on the outskirts of the medieval town, and those upon "made" ground along the western edge of the river terrace where the gravel descends below the alluvium.[1]

[1] T. McKenny Hughes, "The Superficial Deposits of Cambridge and their effect on the distribution of the Colleges", *Proc. Camb. Antiq. Soc.* xi, 293 (1907).

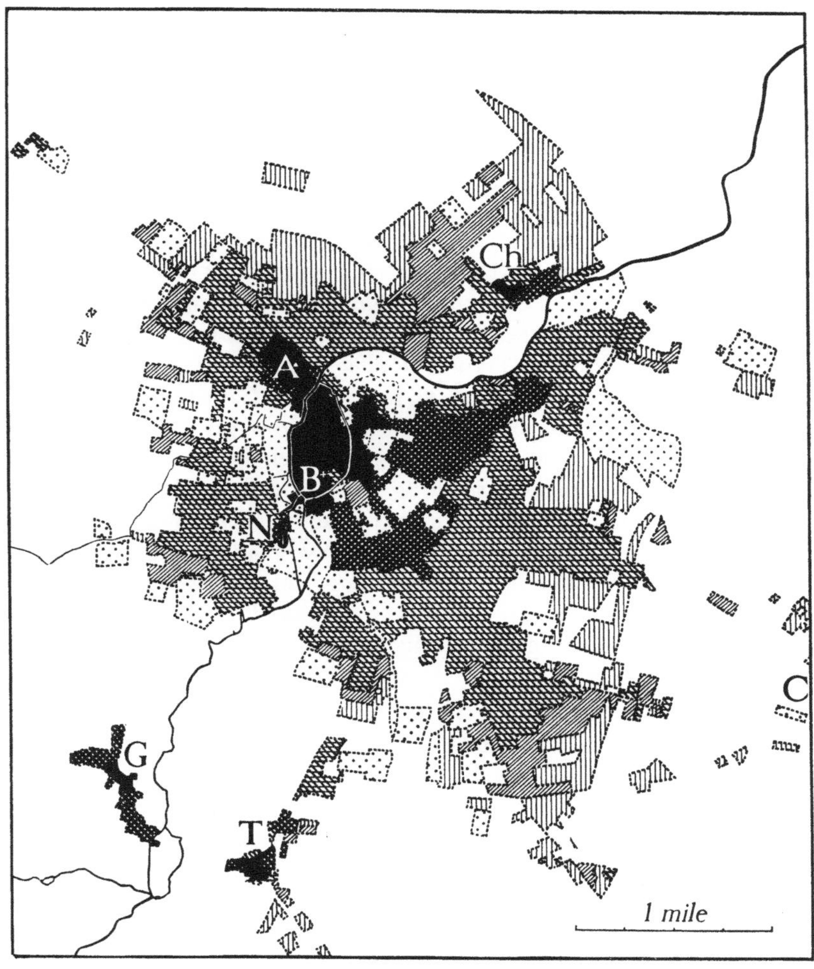


Built-up Area

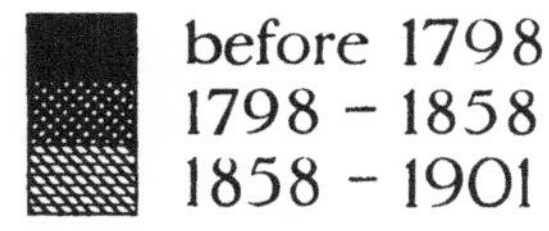

Fig. 39.

The Growth of Cambridge.

A = The Castle. B = St Bene't's Church. N = Newnham. G = Grantchester. T = Trumpington. C = Cherryhinton. Ch = Chesterton. The white line curving from below B to the river opposite A represents the course of the King's Ditch. Commons, etc. includes Commons, the Backs, Playing Fields, Recreation Grounds and Cemeteries. The area shown in solid black was covered with houses by 1574. For sources see footnote 2, p. 168.

The first group of Colleges chose the easier solution; Peterhouse (1284) and Pembroke (1347) were built upon gravels on the southern edge of the town, while the monastic buildings of the Friar Preachers (1240), later utilised by Emmanuel College, indicates the extension of building to the gravels on the east of the King's Ditch depression. Many of the earlier buildings of the second group (Michael House, 1324, Clare Hall, 1326, Gonville Hall, 1348, Trinity Hall, 1350) are clustered around slightly higher ground indicated by the modern name of Senate House Hill, upon which was built the first University buildings, the Grammar School, the Law School, and the Arts School. Lower sites to the north and south were soon utilised, King's Hall (Trinity College) was built in 1337, King's Chapel in 1446, Queens' College in 1448, and St Catharine's College in 1473. As the river Colleges have grown and extended their buildings in modern times, the alluvial river marshes have been drained and raised, and the Cam has been canalised, resulting in the stretch of College gardens, playing fields and commons which constitute the Backs to-day. St John's College, in the nineteenth century, placed new buildings west of the river on a purely alluvial site; but Clare College, in the twentieth century, preferred to separate its new buildings from the old and placed them on the rising ground of the gravel terrace west of the river.

## THE PERIOD 1500–1800

The built-up area of the medieval town can be deduced only indirectly from archaeological and literary sources, but its extent from the later sixteenth century onwards is clearly revealed in the excellent series of plans and maps of varying dates which have survived.[1] The earliest extant of these detailed plans, those of Lyne (1574) and Hammond (1592), have been taken as the basis of the map showing the growth of Cambridge (Fig. 39).[2] These early plans raise complex architectural questions which are not important here: they provide, at any rate, a reliable picture of the extent of the town in the later sixteenth century.

In the south, a few houses flanked the two main roads into the town, Trumpington Street and St Andrew's Street, separated by marshy ground of a depression in the gravels occupied then by St Thomas' Leys and by Swinecroft. This area was known later as the Downing site and was not

[1] J. W. Clark and A. Gray, *Old Plans of Cambridge, 1574–1798* (1921).

[2] Fig. 39 has been constructed from the plans of Richard Lyne, 1574, John Hammond, 1592; the surveys of David Loggan, 1688, William Cunstance, 1798, George Baker, 1830, Richard Rowe, 1858; and the 6 in. editions of the Ordnance Survey of 1885, 1901, 1925. For the extension of the built-up area between 1925 and December 1937, I am indebted to the Cambridge Borough Engineer and Surveyor for permission to use plans in his possession.

built over until the nineteenth century. On the east, Parker's Piece, Christ's Pieces and Butt's Green marked the edge of the built-up area and are shown as cornlands. No buildings of note extended beyond Jesus College; and north of the river the gravel-spread towards Chesterton was entirely open; but houses had crept down the west side of Castle Hill to the edge of the alluvium of the Bin Brook. The Backs are shown as completely rural on the maps; grazing animals on the alluvium suggest meadow, conventional grain fields on the flanking terrace suggest arable land. The river was bridged at Silver Street, but, beyond the bridges, the road was replaced by field paths leading to the small settlement around the Mill at Newnham.

Cambridge at the end of the sixteenth century, then, covered much the same area as the medieval town; the further changes in area, as shown on the plans of Loggan (1688) and Cunstance (1798), are so slight as to be perforce omitted from the Growth of Cambridge map (Fig. 39). These show a few more buildings on the outskirts of the town; but the rural environs as sketched for the sixteenth century remain essentially the same. This almost complete halt in territorial growth during the seventeenth and eighteenth centuries is most striking.

During these centuries, however, there is evidence of a large increase in population within the existing built-up area. The Poll Tax Returns for 1377 record 1902 persons more than fourteen years of age for the Cambridge Borough.[1] Cooper quotes estimates and counts of population out of the Colleges[2] during the early modern period which may be summarised thus:

| 1587 | 1728 | 1749 | 1794 | 1801 |
|---|---|---|---|---|
| 4990 | 6422 | 6131 | 8942 | 9276 |

These figures suggest a considerable increase during the later Middle Ages, followed by a period of slow growth during the seventeenth century changing to relatively rapid growth in the second half of the eighteenth century. The details of the figures for the eighteenth century show that, although there was a general increase of population density, the changes in the central parishes, where the density of population was highest, were not great. A marked increase, however, characterised the parishes with land on the outskirts; the figures for the parish of St Giles show the first

[1] E. Powell, *The East Anglia Rising in 1381* (1896), p. 121.
[2] C. H. Cooper, *op. cit.* ii, 435 (1842); iv, 203, 274, 451, 470 (1852). The number of hearths in the town in 1662 was recorded as 4031. *Ibid.* iii, 501.

sign of growth to the north of the river, later to assume such astonishing proportions.

The growth of the University during this period can best be seen from the Matriculation figures (Fig. 40).[1] These figures fluctuate considerably, but between 1600 and 1675 average 307 per annum; numbers decline during the last quarter of the century, and remain below 200 per annum (average 161) throughout the eighteenth century. This suggests a resident

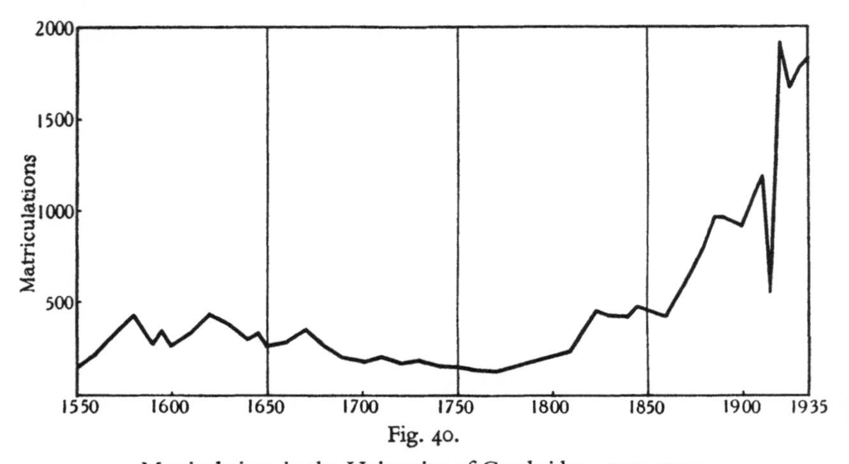


Fig. 40.

Matriculations in the University of Cambridge, 1550–1935.

I am indebted to Dr J. A. Venn for permission to reproduce this graph.

University population of about 1220 in the seventeenth century[2] and 650 in the eighteenth. The curve of Matriculation shows an upward trend in the second half of the eighteenth century and the Census Returns for 1801 record 811 resident members of the University.

The increase of population within the Borough between 1500–1800, without a corresponding increase in the built-up area, indicates therefore a steadily increasing density of population,[3] and the lack of territorial expansion reflects strongly the building restriction caused by the marshes and by the open-fields around the town. The enclosure of the open-fields between 1801 and 1807 was followed at once by a great increase in the built-up area, which represented, in part, the relief from cramped and

[1] C. H. Cooper, *op. cit.* iii, 553, quotes an estimate made by John Ivory in 1672 which gives 2522 as the number resident in the Colleges, including Fellows, Scholars and Servants.

[2] J. A. Venn, "Matriculations at Oxford and Cambridge, 1544–1906", *The Oxford and Cambridge Review*, No. 3, p. 48 (1908).

[3] F. W. Maitland, *Township and Borough* (1898), pp. 101–5.

overcrowded conditions in the old town. The population of the Borough was also increasing rapidly in the nineteenth century (Fig. 41).

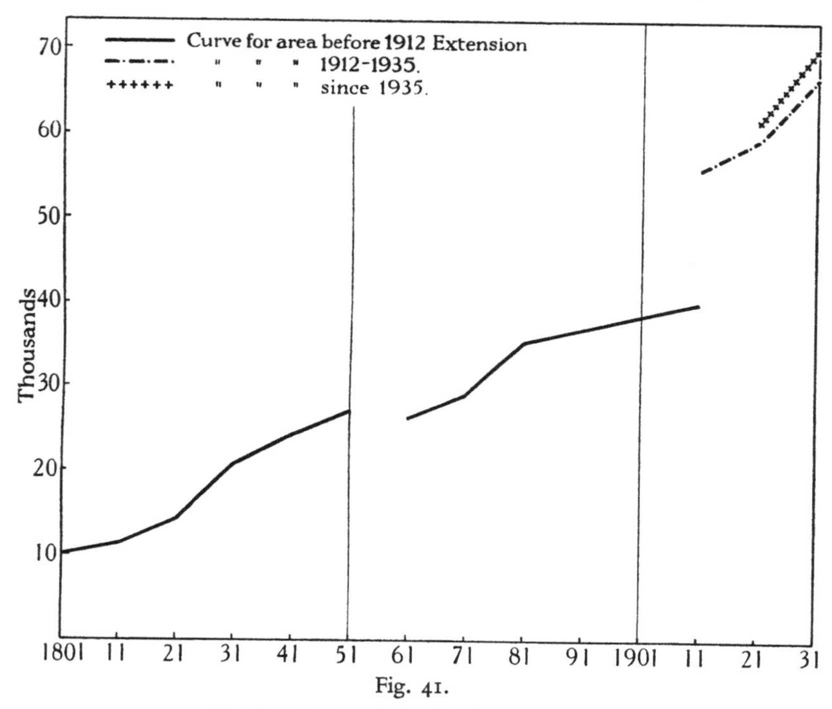


Fig. 41.

The Population of Cambridge, 1801–1931.

The break in the curve between 1851 and 1861 is due to the fact that between 1811–51 the University was in Residence on Census Night, and from 1861–1911 in Vacation. In 1921, Census Night was 19/20 June at the beginning of the Vacation; in 1931 the Census was taken in Full Term. The population in 1921 and 1931, within the area of the Borough as extended in 1935, is shown in addition to that of the area of the Borough at the time of the Census. The official estimate of the population of the Borough in 1936 was 76,760; and in 1938, including the "overspill" beyond the Borough boundary, the population is about 90,000. I am indebted to Dr J. A. Venn for these later figures.

## THE NINETEENTH CENTURY

Fig. 39 shows the area built over between the publication of the plan of William Cunstance in 1798 and that of Richard Rowe in 1858. North of the river there was a small extension to the south-west of Castle Hill, but the striking growth of the town was southward and eastward. A large area on the Intermediate Terrace gravels, on either side of the central depression of the Downing site, was built upon during this period. Beyond the clay strip, largely occupied by the open commons, streets of small houses appear on the western portion of the Higher Terrace. The movement of population to the periphery, and the relief of congestion in the central parishes, are clearly shown by the Census Returns for the nineteenth century (Fig. 42). The 1851 Returns attribute the decrease in Great St Mary's

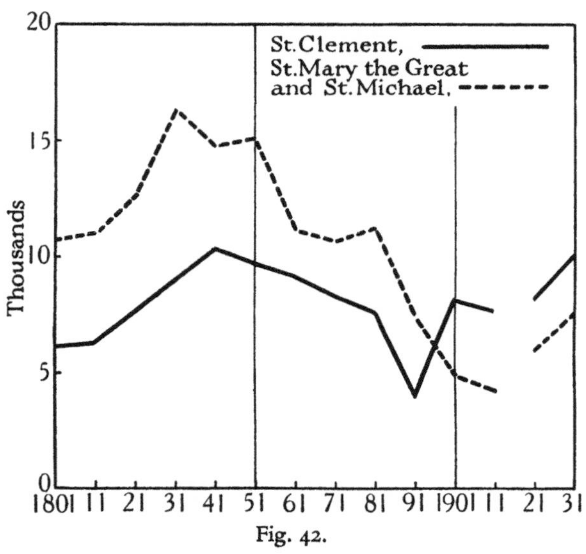


Fig. 42.

Population curves of two central parishes, showing decline in numbers due to movement into the suburbs. (N.B. The break in the curve between 1911 and 1921 indicates a change in the parish boundaries.)

parish to "the recent destruction of houses on Market Hill by fire and not rebuilt"; while, in the parish of the Holy Sepulchre, "the decrease of population is caused by the demolition of a number of old and unsafe tenements". The peripheral growth is seen in the population curve for the parish of St Andrew the Less (Fig. 43), where, says the Census Returns, "the marked increase is due to the erection of public buildings and the enlargement of the Colleges and therefore an increase in the number of labourers and mechanics. Several streets of small houses have been built."

The growth of the town in size and numbers in the first half of the nineteenth century was probably due largely to the progress of medicine and sanitation. Cambridge had suffered in the past repeatedly and severely from pestilence. The increase in University numbers and the building activity of the Colleges added to the prosperity of the town, and there also appears to have been an increase in its importance as a market for the rural areas. The Hay Market in 1820 and the Cattle Market in 1842 were removed from the centre of the town to more spacious sites on Pound Hill (the western slope of Castle Hill). The old Corn Exchange on St Andrew's Hill was also opened in 1842. The outlying villages, Newnham, Grantchester, Trumpington, Cherryhinton, Chesterton, also show in-

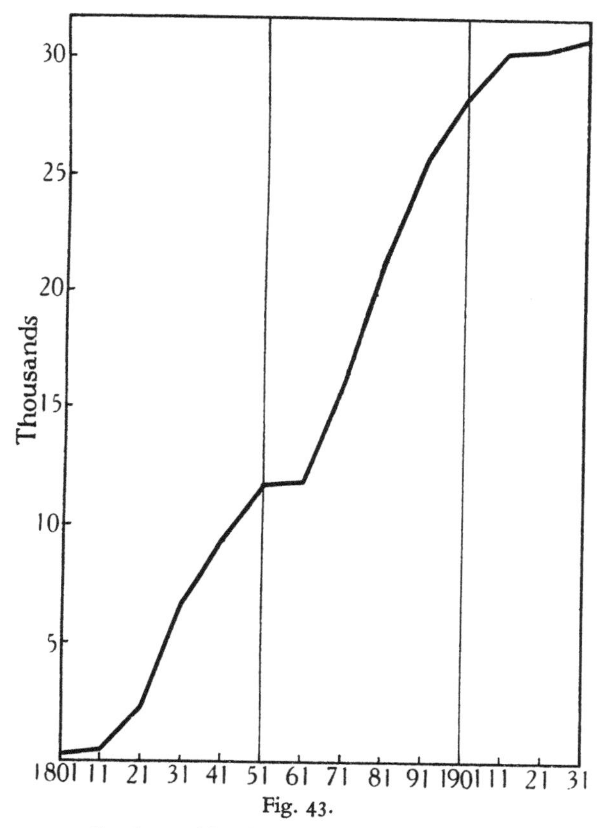


Fig. 43.

Population curve for the parish of St Andrew the Less, including after 1845 the parishes of St Paul (1845), St Matthew (1870), St Barnabas (1888), and St Philip (1903).

crease of population (Fig. 44) and area at this period; in the case of Cherryhinton, the 1821 Census Report specifically attributes the increase to enclosures.[1]

The expansion of Cambridge in the second half of the nineteenth century was even greater (Fig. 39). The town continued to grow rapidly

[1] As was the case with other villages of the County. See p. 129 above.

eastwards upon the gravel forming higher ground between Hobson's Brook and Coldham's Brook. The railway, built along the summit of the ridge, was opened in 1845.[1] The railway fostered development in this area; as early as 1851 the competition of railway transport was causing a shift of population. The riverside parish of St Clement showed a decline in numbers, and the 1851 Census Report declared "many families have left as the Eastern Counties Railway is absorbing the trade of the Cam". Industrial enterprises were attracted to sites near the railway; and the strip of Gault, appearing beneath the gravels in the area between Coldham's Brook

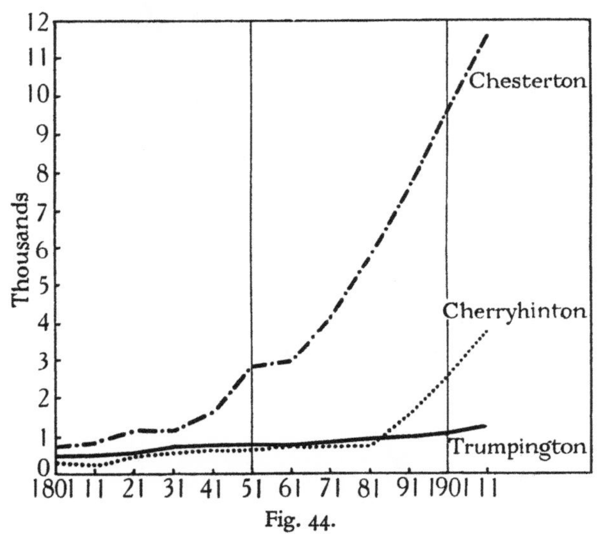


Fig. 44.

Population curves for the outer parishes largely added to the Borough by the Extension Acts of 1911 and 1934 (see Fig. 46).

and the Cam, was utilised for brick and tile works. The parish of St Andrew the Less grew rapidly (Fig. 43), and was subdivided repeatedly during the nineteenth century. Much of this area became a district of mean streets and small, crowded houses, forming in the twentieth century one of the most densely populated areas in Cambridge (Fig. 45).[2]

The other area with a density of more than thirty persons per acre at the beginning of the twentieth century lies to the north of the river on the Intermediate and Lower Terrace gravels which raise the ground above the

[1] See pp. 132–4 above for the opening dates of the various railway lines.

[2] Fig. 45 has been constructed from the Census figures, using the wards before the 1935 changes as the unit areas. These units were chosen in preference to the existing wards because they are smaller, and figures for three decades, 1911, 1921 and 1931, are available.

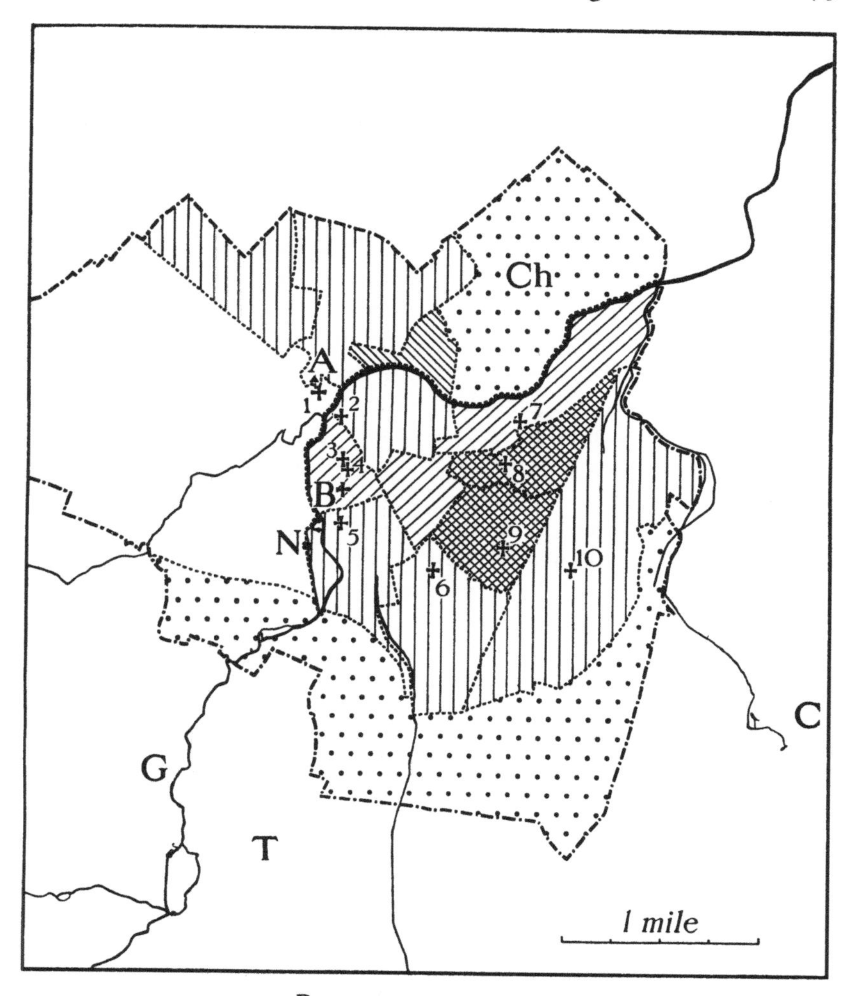


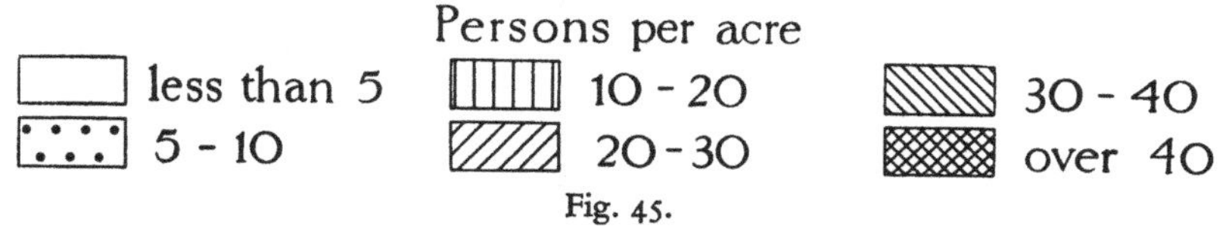


Fig. 45.

Cambridge: Density of Population, 1911–1931.

A = The Castle. B = St Bene't's Church. N = Newnham. G = Grantchester. T = Trumpington. C = Cherryhinton. Ch = Chesterton.

The parish churches marked ✚ are: 1. St Giles'. 2. St Clement's. 3. St Michael's. 4. Great St Mary's. 5. Little St Mary's. 6. St Paul's (New Town). 7. St Andrew's the Less (Barnwell). 8. St Matthew's. 9. St Barnabas' (Petersfield). 10. St Philip's (Romsey Town).

20 ft. contour and so above the flood level. This district was also largely built up, during the second half of the nineteenth century, to provide accommodation for the growing working-class population: indeed, by the end of the century it had stretched out and joined the expanding village of Chesterton. The remarkable growth in the Chesterton district was already foreshadowed in 1851. The Census Returns attribute a decrease in the non-collegiate population of St Giles' to "the removal of many families to the neighbouring suburb of Chesterton"; and the return for Chesterton noted that "upwards of 200 houses have been erected in the last ten years principally inhabited by persons attracted by low rents and light taxation to reside there, though engaged in business in the town of Cambridge". Finally, there was also a considerable expansion of the residential area during this period. Building proceeded apace to the south, along the higher ground on either side of Hobson's Brook, to the west, along the Newnham gravel terrace, and to the north, along the Castle Hill ridge.

Three factors account, in the main, for the growth of the town at this period: (1) the development of railway communication, (2) the development of industries, and (3) the marked growth of the University.

The negotiations between the Town and University authorities took a long and arduous course. A suitable site for a station was very much discussed: that of the Eastern Counties Railway, the first opened, occupied the present site, but it was felt "it was so exceedingly bad that altogether the advantages of the railway were almost superceded by the disadvantage of the station".[1] The Midland and Eastern Company "proposed to remedy that evil" and "to run their line through Coe Fen and bring their station to the very heart of the town". Numerous sites were considered, Sheep's Green, Butt's Green, Midsummer Common among them, but eventually "the difficulties likely to stand in the way of obtaining a site easily accessible and convenient to all the railways likely to branch off from the town and at the same time not interfering with the beautiful walks around the town or with College grounds" proved insuperable. The project for a central station, which might have radically altered the town plan of Cambridge, was abandoned.

It was urged that the railway would "afford unquestionable advantages to a large district hitherto shut out from the benefit of railway communication";[2] and it was argued that "the river would feed the railway and the railway feed the river".[3] The opening of the railway did bring Cambridge

[1] Report of a Railway Meeting, *Cambridge Chronicle*, 22 Nov. 1845. The remaining quotations in this paragraph are also taken from this report.

[2] *Cambridge Chronicle*, 11 Feb. 1843.

[3] *Ibid.* 3 Oct. 1834.

into closer touch with the London market but, in spite of assurances to the contrary, it had a devastating effect on the trade of the Cam; the long lines of barges carrying coal, wood, and stone soon disappeared.

Concurrently with the development of communications came the development of industries. Brick and tile works at Cherryhinton and Coldham's Lane, cement works at Romsey Town, are conveniently placed near the railway. Flour-milling, sausage-making, brewing and malting occupied increasing numbers, and Chivers' jam factory, opened at Histon in 1873, also drew workers from Cambridge. Printing, an old-established industry in the town, occupied 286 men in 1901 and in 1881 the Cambridge Instrument Company was founded. Building and construction work provided employment for a large number of industrial workers.

The two main industries, building and printing, together with retail trade, are in fact closely connected with University development; and the marked expansion of the University, in the second half of the nineteenth century, was the most important single factor in the growth of the town at this period. The numbers of undergraduates rose steadily:

| 1861 | 1871 | 1881 | 1891 | 1901 | 1911 | 1921 | 1931 |
|---|---|---|---|---|---|---|---|
| 1529 | 2097 | 2688 | 3029 | 2958 | 3781 | 4748 | 5204 |

With this increase went a corresponding increase of teaching and administrative officers. After 1871 the abolition of religious tests by the University and Colleges was, at any rate, one among many causes that lay behind the increase. College buildings became inadequate to house the growing numbers, and the demand for lodgings grew. Then, again, in 1882 came the abolition of the rule that Fellowships must be surrendered on marriage; and the same Statutes decreed that Fellowships were to be conditional on active work in the Colleges or the University, thus necessitating residence in Cambridge. These changes could not but affect the growth of the town; married Fellows needed house accommodation as well as rooms in College and this was doubtless a factor in the development of the residential suburbs.

## THE TWENTIETH CENTURY

The twentieth-century extensions of area are shown on Fig. 39 for two periods. The first quarter of the century is differentiated from the development of the last ten years in order to emphasise present tendencies. The period 1901–25 was characterised by rapid development along the main roads; the last ten years by an attempt to control ribbon development and to fill in the empty areas between these roads.

In 1906 it was said that "our town is mostly built";[1] even in 1925 it could be stated "Cambridge is still to an appreciable extent a rural township".[2] Present development is rapidly changing these conditions. Twice at short intervals, in 1912 and 1935, the Borough boundaries (Fig. 46) have been considerably extended—boundaries hitherto unchanged throughout the centuries. The additions to the Borough show the position and amount of the extension:

*By Extension Order Act of* 1911.

| Parish | Area | Population |
|---|---|---|
| Chesterton (whole) | 1,173 acres | 11,330 persons |
| Cherryhinton (part) | 388 | 2,749 |
| Trumpington (part) | 497 | 527 |
| Grantchester (part) | 166 | 1,179 |

*By Extension Order Act of* 1934.

| Parish | Area | Population |
|---|---|---|
| Cherryhinton (whole) | 1,671 acres | 1,254 persons |
| Trumpington (part) | 1,439 | 1,179 |
| Gt. Shelford (part) | 188 | 74 |
| Fen Ditton (part) | 441 | 437 |
| Impington (part) | 570 | 341 |
| Milton (part) | 294 | 95 |

The most rapid growth is now in the south, and along the gravels to the north of the river. The tendency to move out from the medieval nucleus has largely ceased and the recent extension of College buildings will result, if anything, in a rise of population density in this area during term time.

But the general movement to the periphery continues. The residential areas to the south and west are still growing. There is also a marked movement from the eastern and northern slums of the nineteenth century to the adjoining areas. The wards of South Chesterton, Petersfield and St Matthew, which had a density of more than 40 persons per acre in 1911, showed a decrease to 37·3, 38·8, and 37·8 persons per acre respectively in 1931. During the same period the density increased in the rest of Chesterton and in Romsey Town (Fig. 45). A new factor is at work here, the twentieth century demands better standards in housing conditions and looks askance at the crowded buildings of the previous one. This factor is additional to the continued development of the University, and to the continued increase in industries, largely of a skilled character: the Cambridge

[1] E. M. Jebb, *Cambridge. A Brief Study of Social Questions* (1906), p. 25.
[2] A. Gray, *The Town of Cambridge* (1925), p. 167.

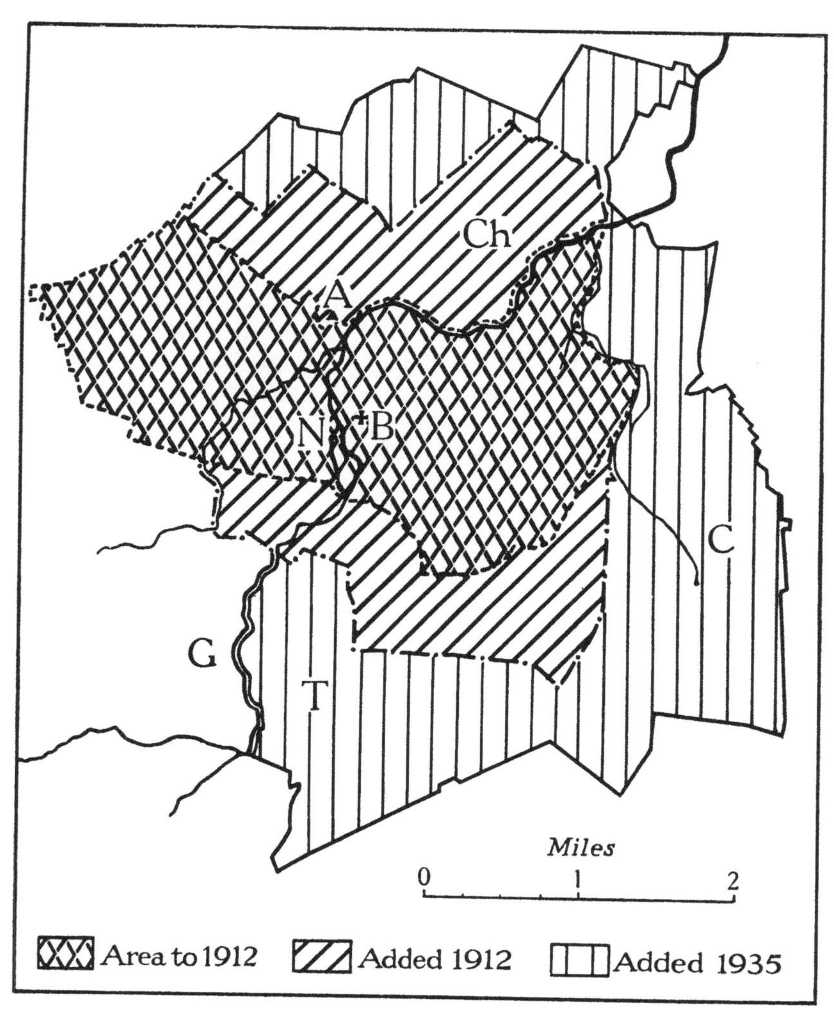


Fig. 46.

Cambridge: Extensions of the Borough.

The Borough boundaries are taken from the town plans of Cambridge published by Mr W. P. Spalding. The Extension Order Acts are dated 1911 and 1934; they came into effect in 1912 and 1935.

A = The Castle. B = St Bene't's Church. N = Newnham. G = Grantchester. T = Trumpington. C = Cherryhinton. Ch = Chesterton.

Instrument Company now employs 700 hands and the Pye Radio Works is a new and important industry.[1]

The *laissez-faire* development of the nineteenth century has been taken in hand. The Town Planning Department is in full working order, and the Borough land has been tentatively allotted to various purposes. Building schemes of twelve houses to the acre are a feature of the east and north-east, to which area, with its good railway sites, it is hoped to confine new and expanding industrial enterprises. In the west it is planned to curtail the density of building to, at most, four houses to the acre.

There was a proposal of the Town Council in 1841 to enclose portions of the Commons for building sites and market gardens.[2] Fortunately for the beauty of Cambridge and the preservation of the individuality of the old town, this was turned down by a meeting of the townsmen "characterised by extreme noise and tumult".[3] Medieval Cambridge is thus largely separated from the expanding Cambridge of to-day by a ring of open land formed by the Commons and the Backs.

[1] See pp. 159–60 above.
[2] Report read to a meeting of the Town Council, April 1841. Quoted by C. H. Cooper, *op. cit.* iv, 633.
[3] C. H. Cooper, *op. cit.* iv, 634.

## Bibliographical Note


(1) C. H. Cooper, *Annals of Cambridge*, 5 vols. (1842–53).
(2) R. Willis and J. W. Clark, *The Architectural History of the University of Cambridge*, 4 vols. (1886).
(3) J. B. Mullinger, *The University of Cambridge*, 3 vols. (1873–1911).
(4) F. W. Maitland, *Township and Borough* (1898).
(5) F. W. Maitland and M. Bateson, *The Charters of the Borough of Cambridge* (1901).
(6) A. Gray, *The Town of Cambridge* (1925).
(7) A. Gray, *Cambridge University. An Episodical History* (1926).

CHAPTER THIRTEEN

# THE DRAINING OF THE FENS A.D. 1600–1850

By H. C. Darby, M.A., PH.D.

DURING THE MIDDLE AGES THE DRAINING OF THE 1300 square miles of the Fenland had remained largely a matter for local concern. When necessity arose, owing to the ravages of the sea or to the overflowing of the watercourses, the Crown granted a commission to remedy the evil. A succession of Commissions of Sewers combined with local custom to maintain the medieval economy. The upkeep of any single channel involved many interlocking interests, and the dissolution of the monasteries in 1539 served only to increase the confusion of divided responsibilities. But, as Samuel Hartlib wrote, "in Queen Elizabeth's dayes, Ingenuites, Curiosities and Good Husbandry began to take place". The time was becoming ripe for a "greate designe" in the Fenland. During the later years of the sixteenth century, various schemes and experiments prepared the way. At last, in 1600, there was passed "An Act for the recovering of many hundred thousand Acres of Marshes...". Of the many stretches of marsh in the kingdom, that of the great Fenland itself provided the most spectacular transformation.

Many schemes were afoot during the early years of the seventeenth century, and there was great opposition from those with vested interests in the fen commons and in the fenland streams. There was also much debate about ways and means. Nothing effective was done; general dissatisfaction was felt everywhere. The net result was that some fenmen approached Francis, 4th Earl of Bedford, the owner of 20,000 acres near Thorney and Whittlesea, who contracted within six years to make "good summer land"[1] all that expanse of peat in the southern Fenland, later known as the Bedford Level. An agreement was drawn up in 1630. In the following year, thirteen Co-Adventurers[2] associated themselves with the earl; and in 1634 they were granted a charter of incorporation. Their hope was to turn this expanse of "great waters and a few reeds" into "pleasant pastures of cattle and kyne"; and they secured the services of the Dutch engineer Vermuyden, who had been at work upon the reclamation

[1] I.e. Land free from floods in summer. This is the story told by C. Vermuyden in *A Discourse touching the drayning the great Fennes* (1642).

[2] So called because they "adventured" their capital. See p. 105 above.

of the Axholme marshes. Under his direction, cuts, drains, and sluices were made. Chief among these was the Old Bedford River extending from Earith to Salter's Lode, 70 ft. wide and 21 miles in length.[1]

In 1637, at a Session of Sewers in St Ives, the Level was judged to have been drained according to the true intent of the agreement of 1630. But complaints and petitions showered upon the Privy Council, and royal feeling turned against the Corporation. The inner history of this change in royal favour is obscure; at any rate, in the following year, the award was set aside. The Level still remained subject to inundation in winter, and so it was maintained that the contract of 1630 had not been fulfilled. The king himself, now, planned to drain the Fens "in such manner as to make them winter grounds", and he retained the services of Vermuyden. Soon, however, the fen difficulties were overshadowed by greater troubles. The country was at war within itself.

During the Civil War the draining was in abeyance, but the project had not been forgotten. After many committees and sub-committees, an "Act for the draining the Great Level of the Fens" was passed in May 1649; and the 5th Earl of Bedford and his associates were "declared to be the undertakers of the said work". In his *Discourse* of 1642, Vermuyden had divided the Great Level into three areas:

"1. The one from Glean to Morton's Leame.
2. From Morton's Leame to Bedford River.
3. From Bedford River southwards, being the remainder of the level."

These became the North,[2] the Middle, and the South Levels respectively (see Fig. 47). Despite continued hostility, activity was restarted under Vermuyden. The earlier works were restored; banks were made; sluices built; and channels scoured. In particular, the New Bedford River was cut running parallel to the Old Bedford River.[3] Between the two Bedford Rivers, a strip of land[4] was left open to form a reservoir for surplus water in time of flood (see Fig. 48). The old course of the Ouse was sluiced at Earith (the Hermitage Sluice) and at Denver, so that it became merely the drain for the fens in the Isle of Ely. The New Bedford

[1] Before this, the Old West River carried part of the Ouse in a circular course around the Isle of Ely and so to Denver, and thence to the sea at Lynn. Now, this water reached Denver directly through the Old Bedford River.

[2] Later, however, the North Level did not extend beyond the Welland.

[3] The New Bedford River was alternatively known as the Hundred Foot River, and, for the greater part of its course, ran half a mile to the east of the older cut.

[4] This became known as "The Wash", "The Washes", or "The Washlands". High "barrier" banks on the outer sides of the two cuts kept the water within definite limits. It could be run off at convenience; in the early part of the nineteenth century Welmore Lake Sluice was built to facilitate the run-off into the New Bedford River above Denver Sluice.

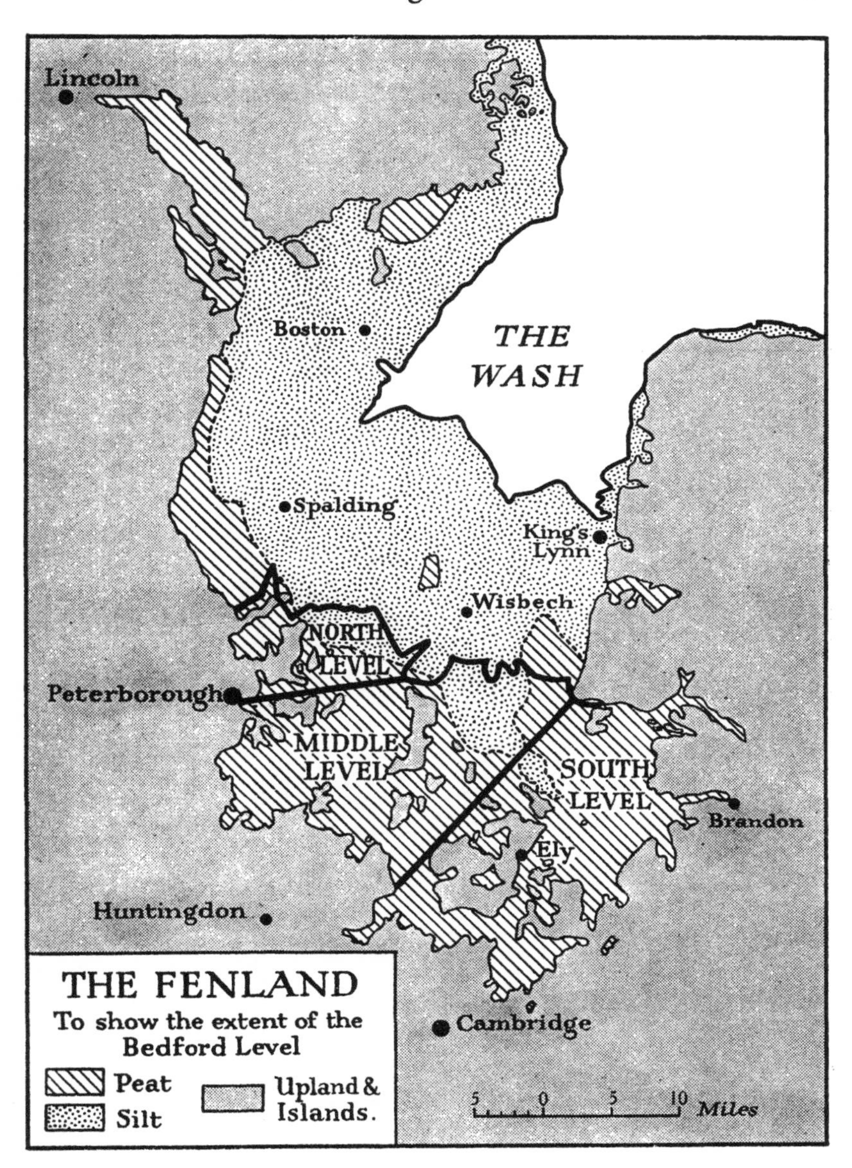


Fig. 47.

S. B. J. Skertchly, in *The Geology of the Fenland* (1877), p. 129, noted that the precise boundaries of the peat and silt were "very obscure, for the peat thins out insensibly along its borders." The limits of the Bedford Levels are taken from Samuel Wells' map of 1829 on a scale of $1\frac{1}{2}$ miles = 1 inch.

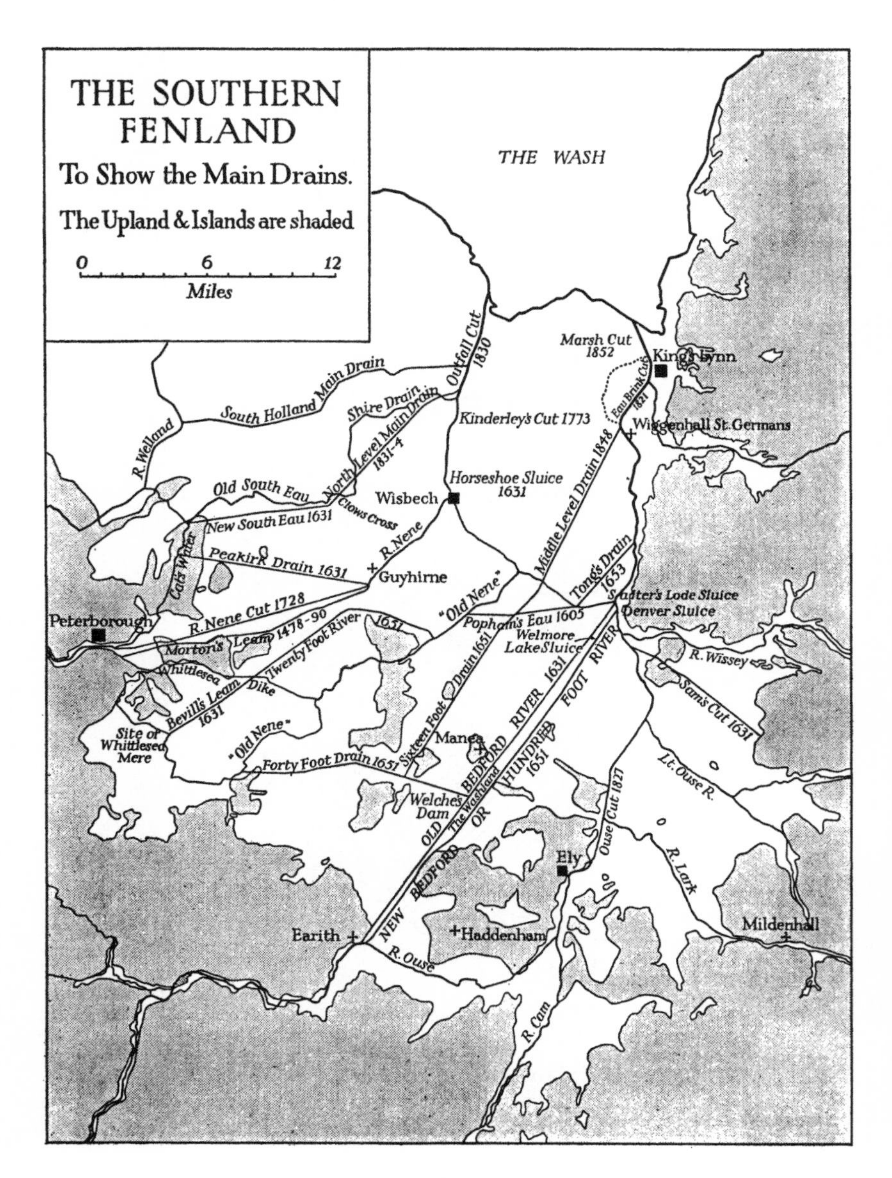


Fig. 48.

The approximate dates of the drains are given. In some cases there was an appreciable interval between the start of a project and its completion.

River became the main channel of the Ouse. The Seven Holes Sluice at Earith[1] kept the waters of the Ouse from flowing into the Old Bedford River (see Fig. 49). The final warrant of adjudication came in March 1652. Successive changes in the administration of the realm witnessed the completion of the machinery for preserving the works of the drainers, until, finally, there was passed the General Drainage Act of 1663.

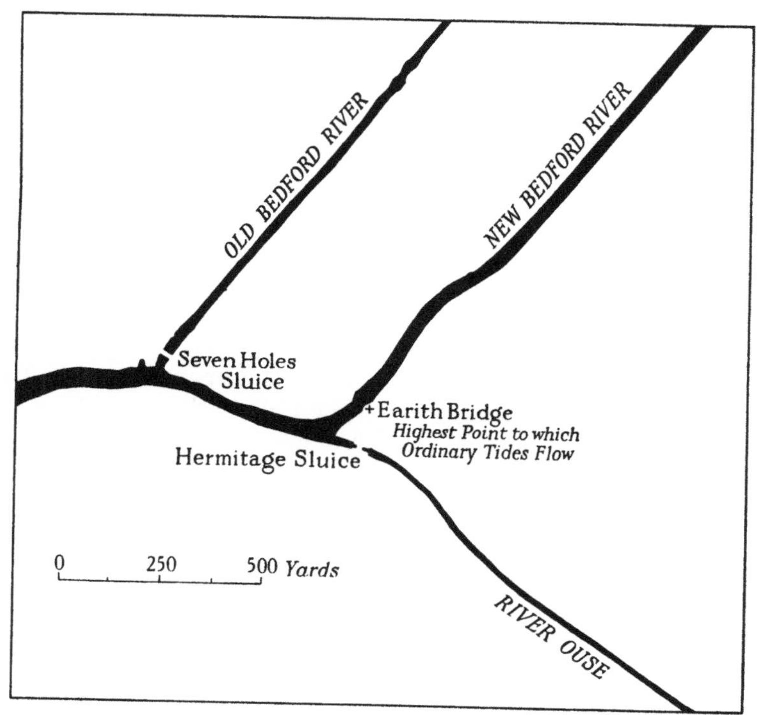


Fig. 49.

Seven Holes Sluice and Hermitage Sluice.

The highest point to which high spring tides flow is at Brownshill Staunch, some 2¼ miles above Earith Bridge.

At first, great success followed upon the works of the drainers. Cultivation was introduced on land that, as far as record went, had never before known a plough. As Thomas Fuller wrote in 1655, "the best argument to prove that a thing may be done is actually to do it". But time was not to fulfil these hopes. Despite many praises, it soon became evident that all

[1] Thus it protects the Washlands. But if the water coming down the upland Ouse is considerable it is opened—in summer rarely, in winter more often.

was not well in the Bedford Level. Some of the complaints that followed the final adjudication in 1652 were only to be expected—disputes about the allotment of the reclaimed land, and about the management of the new drains. These were problems of routine administration; they could be settled by negotiation and compromise. But there were other difficulties of a much more fundamental character, difficulties that brought the very success of the drainers near to disaster. Right up to the present day these difficulties have remained important in all discussions about draining. They are of two kinds.

## THE LOWERING OF THE LEVEL OF THE FENS

The first group of difficulties resulted from the drying up of the peat fen. As the peat was drained it rapidly became lower in level. This lowering was due in part to the shrinkage of the peat, and in part to the wasting away of the peat surface owing to bacterial action and owing to cultivation. As a result, the surface of the peat soon became lower than the level of the channels into which it drained. The channel beds were lined with silt, and so escaped as rapid a lowering. This difference in height can be seen to-day along many of the fen rivers; they are at a higher level than the land through which they flow. The small drains right in the heart of the peat area suffered most. In time, they came to flow at a lower level than the main cuts into which they tried to discharge their waters! And the more these evils were combated by more effective draining, the more rapidly the peat surface continued to sink. Thus it was that the works of one generation became inadequate for the needs of the next.

An idea of the amount of this lowering can be seen to-day from Holme post in Huntingdonshire just outside the boundary of Cambridgeshire, along what was the south-western margin of Whittlesea Mere. In 1851, the top of this iron column was even with the surface of the ground. By 1870, nearly 8 ft. was exposed. To-day, it stands about 11 ft. high. This case is fairly extreme. The amount of shrinkage in any particular locality depends upon the original thickness of the peat as well as upon the intensity of the drainage operations. Even a shrinkage of half an inch per annum is important. It may be viewed with equanimity from one year to another, but the result over a period of years becomes critical.

Further, not all the Fenland is peat. This only made matters worse. The coastal areas are composed of silt (see Fig. 47), less liable to shrinkage and wastage than peat. Before the draining, the silt zone was over 5 ft. *lower* than the peat area that lay inland. To-day, the silt area is about 10 ft. *higher* than the peat. As a result of this differential shrinkage, the beds of the outfall channels became almost as high as the peat fen behind.[1]

[1] See p. 191 below.

The consequences of these fundamental difficulties were apparent even before the seventeenth century was over. Soon, disaster was abroad everywhere. What had seemed a promising enterprise in 1652 had become a tragedy by 1700. There was but one way to save the situation—the substitution of an artificial for a natural drainage. Water was pumped out from small dyke to drain, from drain to river, and so to sea.

An early mode of fen drainage was the horse mill, but the only satisfactory source of power at hand was the wind. The introduction of windmills for pumping purposes was, in fact, the critical factor that saved most of the Fens from being re-inundated. As the seventeenth century passed into the eighteenth, windmill drainage became more and more frequent. The whole of the Fenland came to consist almost entirely of small sub-districts, each pumping its water into one of the larger drains that traversed the region. For example, a pamphlet of 1748, written by Thomas Neale, stated that there were no less than 250 windmills in the Middle Level. "In Whittlesey parish alone, I was told by some of the principal inhabitants there are more than fifty mills, and there are, I believe, as many in Donnington (*sic*) with its members. I myself, riding very late from Ramsey to Holme, about six miles across the Fens, counted forty in my view."

But the windmill was far from being the perfect engine. It was at the mercy of gale and frost and calm. It was never very powerful, and soon it ceased to provide a satisfactory solution to the problem of clearing water from the drains. For as the surface level continued to subside, the windmill became increasingly ineffective. Inundations grew frequent. It is easier to put down statistics relating to these "drownings" than to imagine the bankruptcy and distress when crops not merely failed but completely disappeared beneath the rising waters. By the end of the eighteenth century, according to Arthur Young,[1] there were many fens "all waste and water", where twenty years previously there had been "buildings, farmers and cultivation". Some places had been particularly unfortunate: "three years ago five quarters of corn an acre; now sedge and rushes, frogs and bitterns". It was with dismay that he viewed the scene spread before him in the summer of 1805:

> It was a melancholy examination I took of the country between Whittlesea and March, the middle of July, in all which tract of ten miles, usually under great crops of cole, oats and wheat, there was nothing to be seen but desolation, with here and there a crop of oats or barley, sown so late that they can come to nothing.

He predicted the ruin of the whole flat district.

> The fens are now in a moment of balancing their fate; should a great flood come within two or three years, for want of an improved outfall, the whole country, fertile as it naturally is, will be abandoned.

[1] A. Young, *Annals of Agriculture*, xliii, 539 et seq. (1805).

Other evidence bears out the impression of desolation.[1] As one traveller of 1833 could write: "We are now in the very perfection of the fen-country, being several feet below the level of the great running streams, upon land subject to frequent inundation."

In addition to land that had deteriorated, some patches of original fen remained; in the west, for example, were the large reed-bordered lakes of Whittlesea Mere and Ramsey Mere. The great copper butterfly was not yet extinct; nor were all the species of fen birds; nor yet was the ague against which the fenmen took their opium pills. Indeed, many people were still "fearful of entering the fens of Cambridgeshire lest the Marsh Miasma should shorten their lives". That was in 1827. By 1858, the complaint had become "infrequent". The improvement was generally ascribed to better drainage.

Not only malaria, but many other distinctive features of the Fens disappeared before the changes of the nineteenth century. The time came when "sportsmen from the University" were no longer able to indulge a passion for shooting in the fens of Teversham, Quy, Bottisham, and Swaffham. And, in 1854, Henry Gunning was "happy to say that these incentives to idleness no longer exist. Thousands and tens of thousands of acres of land, which at the time I speak of produced to the owners only turf and sedge, are now bearing most luxuriant crops of corn."

The important factor that was giving the fen country of the nineteenth century this more stable economy was the advent of the steam-engine. The possibility of steam-driven pumps for draining had been discussed before 1800, but the idea was slow in gaining support. At length, John Rennie induced the proprietors of Bottisham Fen to erect a small engine to help their windmills.[2] There, in 1820, the first Watt engine was applied to work a scoop-wheel. Despite predictions of failure, other steam-engines followed and soon justified their introduction. An inscription on a pumping station along the New Bedford River is dated 1830, and reads:

> These fens have oftimes been by water drowned,
> Science a remedy in water found,
> The power of steam she said shall be employed,
> And the Destroyer by Itself destroyed.

It was a premature claim, but by 1838 Joseph Glynn, one of the pioneers of steam pumping, certainly had "the pleasure to see abundant crops of wheat take the place of the sedge and the bulrush". The "swamp of marsh, exhaling malaria, disease and death" had been converted into

[1] See p. 117 above. [2] See p. 120 above.

"fruitful cornfields and verdant pastures". By the middle of the century, according to one estimate, the number of steam-driven pumps between Cambridge and Lincoln was about 64; the number of windmills had declined from about 700 to about 220.

Further improvement was at hand. Among the great sights that "astonished the visitors" to the Great Exhibition of 1851 was Appold's centrifugal pump; and its application to fen problems was immediately realised. One of the new Appold pumps was erected to drain Whittlesea Mere. And so, witnessed by "large crowds of people", there disappeared the last remaining large stretch of water in the Fenland—a stretch of water that had enjoyed considerable reputation as the scene of regattas in summer and of skating in winter. The wind which, "in the autumn of 1851 was curling over the blue water of the lake, in the autumn of 1853 was blowing in the same place over fields of yellow corn".[1]

The success of the steam-engine in the Fenland did not mean that all difficulties were over. The lowering of the peat surface necessitated a constant building-up of the river banks. In the absence of easily accessible clay, many banks had been made of peat or light earth, and, during floods, they were subjected to considerable hydrostatic pressure. Breaches were frequent. The paradox was that an effective draining only increased the lowering of the peat surface. What this meant in terms of pumping can be seen from a solitary example. Methwold Fen, until 1883, had drained naturally into the Ouse through a dyke, Sam's Cut.[2] In that year, owing to the lowering of the fen, artificial drainage became necessary, and a pump was erected. During the years that followed the fen continued to sink so rapidly that a second pump had to be installed in 1913. Then, it was estimated that the surface had sunk "5 to 6 feet within the last 50 years". This one example illustrates conditions generally. It was the steam-engine that turned the desolation of 1800 into some prospect of prosperity.

## OUTFALL PROBLEMS

The second group of changes that marked the nineteenth century was associated with the outfalls of the fen rivers into the Wash. In a normal river, the current of water is strong enough to force its way out to sea. But the fenland rivers were far from normal. The downward force of the fresh waters in the gently graded streams was no match, especially in summer, for the strong tidal flow twice each day. With swift flood tides and weak ebb tides (see Figs. 50 and 51), deposition was inevitable, and the

[1] W. Wells, "The Drainage of Whittlesea Mere", *Jour. Roy. Agric. Soc.* (1860), pp. 140–1.

[2] See p. 196 below.

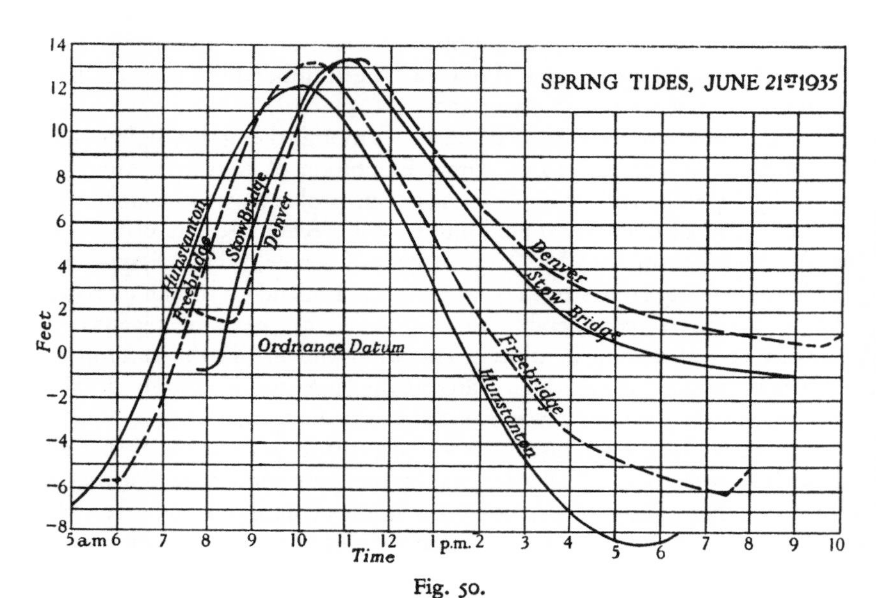


Fig. 50.

Tidal Curves for the Great Ouse Outfall.

Reproduced by the courtesy of the Chief Engineer of the River Great Ouse Catchment Board.

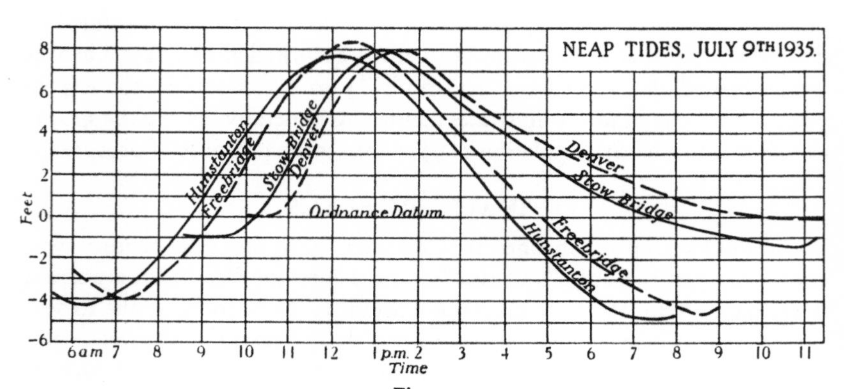


Fig. 51.

Tidal Curves for the Great Ouse Outfall.

Reproduced by the courtesy of the Chief Engineer of the River Great Ouse Catchment Board.

beds of the estuaries were continually being raised. This fact, combined with the lowering of the peat surface in the interior of the Fenland (see Fig. 47), made the outfall channels almost as high as the peatlands behind.[1] The fresh waters found great and greater difficulty in reaching the sea, and the maintenance of a clear channel seawards became of paramount importance. As Colonel Dodson wrote in 1664, just after Vermuyden's draining: "if we cannot be masters there, all other endeavours signifie nothing". The great controversies of the eighteenth century were outfall controversies, concerned with topics like the disposition of sluices, the mechanics of silting, the formation of sandbanks, and the nature of tidal scour. Thomas Badeslade's pamphlet of 1729 was but one of very many. As he said:

all parties acknowledge the misfortune, for they all suffer; but all do not agree in the cause of this general calamity, nor in the method that must be put in practice to relieve them; but all agree and declare, that if something be not done, this country will be rendered uninhabitable.

By 1800, the Ouse reached the sea through a channel of varying width, filled with shifting sandbanks; in the Nene, "ships of large burden could no longer reach" Wisbech; the Welland estuary, too, was full of shoals; that of the Witham was in no better plight. During the nineteenth century, various remedies were tried. New cuts were made, straightening and improving the lower tidal courses of the rivers. The estuary most affecting the Cambridgeshire fens was that of the River Ouse. Above King's Lynn, the Ouse (carrying so much of the water of the Bedford Level) made an extensive bend of about 6 miles to St Germans (see Fig. 48). The channel was nearly a mile wide in places, and comprised a number of uncertain streams. During floods, the flow of the river was much impeded, and it was clear that no improvement could result until the obstructions had been cleared. The remedy was to cut off the great bend of the Ouse; to make the new channel large enough to contain the whole body of the river; and, incidentally, to increase the velocity of the current by shortening the line of the stream.[2] An Act was obtained in 1795 enabling the cut to be made, but the work was not started until 1817. Finally, in 1821, seventy years after it had been first proposed, the Eau Brink Cut was opened. It had an immediately beneficial effect upon the Middle and South Levels—until further accumulations of silt made new improvements urgent. In 1852, further straightening was secured by means of the Marsh Cut towards the sea. Some years before this (in 1846), the Norfolk Estuary Company had been established. Intended to recover land from the Wash, the primary object of this body became the maintenance of an extension seawards of the Eau Brink Cut. Training walls were built to induce the

[1] See p. 186 above. [2] See p. 117 above.

River Ouse to keep a defined channel among the shifting shoals that, accumulating in the estuary, interfered with the effective discharge of

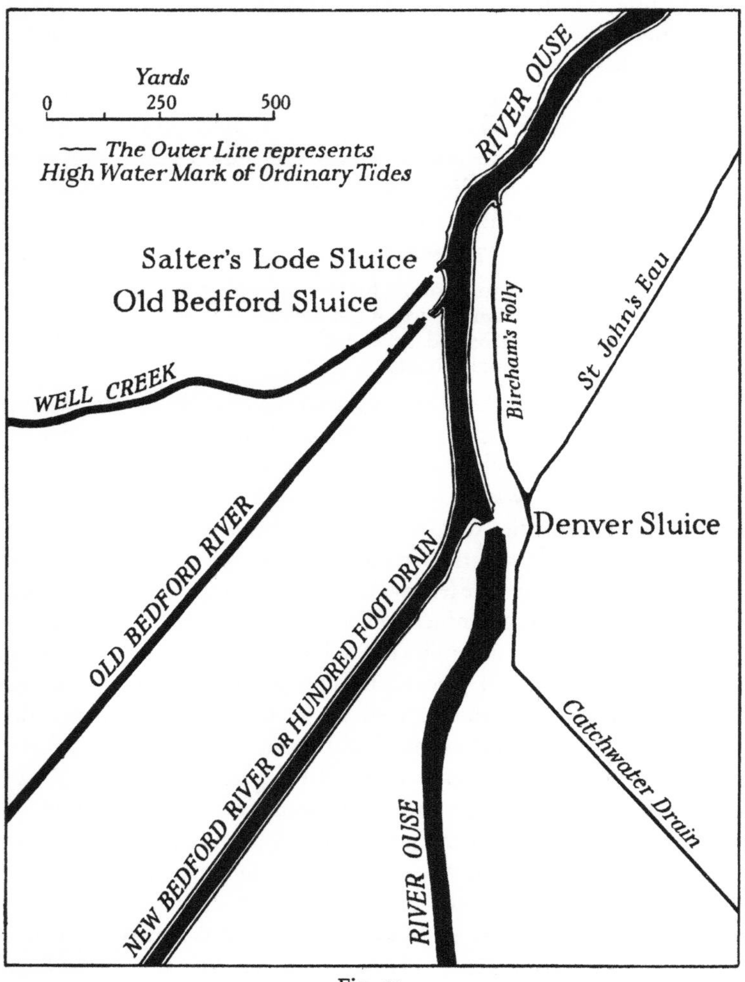


Fig. 52.
The Sluices near Denver.

water from the two Levels.[1] But the nineteenth century brought no solution to this problem.

Important features of the outfalls were the sluices necessary to prevent the tidal waters from passing up the rivers. On the Witham, were

[1] See p. 201 below.

the Grand Sluice and the Black Sluice; on the Nene, the North Level Sluice; on the Ouse, the St Germans Sluice and Denver Sluice.[1] Upon Denver Sluice depended the safety of the South Level. Its importance will be apparent from E. G. Crocker's summary in 1913: "The level of the top of the banks is from 12 to 13 feet above O.D. whilst an ordinary spring tide rises to 14 feet above O.D. at the sluice, the highest recorded tide being 17·51 feet above O.D., so that should anything occur to prevent these gates closing in a spring tide, practically the whole of the Fens of the South Level would be flooded." There was a further complication. Immediately below Denver Sluice, the waters of the Hundred Foot River (carrying the upland Ouse from Earith) fell directly into the estuary (see Fig. 52). Consequently, when there was a great volume of upland water passing down the Hundred Foot River, the level of water in the Ouse outfall (on the seaward side of the sluices) never fell low enough for the sluicegates to be opened to provide an adequate run-off for the waters of the South Level. These waters could only accumulate within the straining banks of their dykes and drains. A crisis was, therefore, always liable to be produced by the combination of (1) adverse wind conditions, (2) a high spring tide and (3) heavy land floods.

## Bibliographical Note

The older works of most general interest are:

(1) S. Wells, *The History of the Bedford Level*, 2 vols. (1830).
(2) S. B. J. Skertchly, *The Geology of the Fenland* (1877).
(3) S. H. Miller and S. B. J. Skertchly, *The Fenland Past and Present* (1878).
(4) S. Smiles, *Lives of the Engineers* (1st ed. 1861), has much interesting material.
(5) Sir William Dugdale's *History of Imbanking and Drayning* (1662). This is, of course, the classic account of the seventeenth-century draining.

The following more recent accounts give detailed sources for the facts recorded in this chapter:

(6) H. C. Darby, "Windmill Drainage in the Bedford Level", Official Circular No. 125, Brit. Waterworks Assoc. (1935); also *The Engineer*, clx, 75 (1935).
(7) H. C. Darby, "The Draining of the Fens A.D. 1600–1800" in *An Historical Geography of England before A.D. 1880* (1936).
(8) H. C. Darby and P. M. Ramsden, "The Middle Level of the Fens and its Reclamation", *Victoria County History of Huntingdon*, iii, 249 (1936).

[1] In addition to the sluices at the outfalls of the Old Bedford River and Well Creek (see Fig. 52).

CHAPTER FOURTEEN

# MODERN DRAINAGE PROBLEMS: 1850–1938

By Oscar Borer, B.E. (N.Z.), M.INST.C.E., A.M.I.MECH.E.
*Chief Engineer, The River Great Ouse Catchment Board*

DESPITE THE ADVENT OF THE STEAM-ENGINE IN THE nineteenth century, it was reported that windmills were still used in parts of Norfolk in 1913; indeed, at Soham Mere, in Cambridgeshire, a windmill still supplements the steam plant. But these are exceptions. In general, the last hundred years have witnessed many changes in fen pumping. After the introduction of the centrifugal pump in 1851, the scoop-wheel was gradually discarded, but not before its diameter had in many cases increased to 36 ft., and even to 50 ft., to accommodate the lowering surface of the land.[1] Its efficiency, however, was always low, being in the region of about 30 per cent. Still, scoop-wheels were old friends, and with all their splashing they had handled large quantities of water against low heads.

There have also been other changes in the steam-driven plants. Gradually, more modern types replaced the beam and the old low-pressure steam-engine.[2] Thus at Prickwillow (near Ely) a new pumping engine was installed in 1897 to replace a side-lever condensing engine of 60 nominal horse-power. This had been erected in 1833; it had used steam at a pressure of 6 lb. per sq. in.; and it had driven a scoop-wheel 33 ft. 6 in. in diameter. The capacities of the old and new machinery are of interest:

| Type of plant | Date | Steam pressure (lb. per sq. in.) | Revolutions of engine (per minute) | Lift in ft. | Lift in in. | Water lifted (tons per minute) |
|---|---|---|---|---|---|---|
| Side-lever condensing engine driving scoop-wheel | 1853 | 6 | 25.5 | 9 | 10¾ | 68·9 |
| Vertical compound condensing marine engine direct-driving horizontal centrifugal case-pump | 1897 | 76 | 132 | 13 | 4 | 153·2 |

[1] See p. 186 above.

[2] For a description of some of the older types of engines, see R. W. Allen, "Modern Pumping Machinery for Drainage of the Fens", *Proc. Inst. Mech. Engin.* (1913), p. 787.

The increase in lift, owing to the lowering surface, stands out. By 1913, this had become 15 ft. 4 in.—or an increase of nearly 6 ft. over a period of 80 years. Some of the older engines still exist to-day, standing near the modern pumps. There is one at Upware; another (installed about 1840) stands alone in the Glassmore district of the Middle Level.

Crude oil or Diesel engines were introduced for pumping purposes about the year 1913. The weight per unit horse-power is much less than for steam—a great advantage in an area of soft earth where foundations are expensive. There are, however, other advantages of special moment. Most of the pumping stations are, of necessity, near river banks and hence away from hard roads. The cartage of coal, particularly if extra supplies are required during winter, is expensive. But the great loss with steam comes in starting-up and stand-by costs. Pumping for drainage is mainly seasonal, and even then spasmodic. With heavy rainfalls, the pumps must be under way by the time the water has percolated into the drains. When this water has been discharged from drain to river, the engine must ease down, or even stop, for some hours until the drains fill up again. With a steam-driven plant, steam must first be raised in the boilers in anticipation of pumping; then, when the pump has shut down, the fires must be either banked or drawn until the plant is required again. This involves unnecessary fuel consumption and may at times require the services of an extra driver. On the other hand, the use of oil-driven plants does mean that, in the event of hostilities, provision will have to be made for a supply of fuel-oil. It is possible, therefore, that the steam-engine could be kept, with advantage, as a stand-by, and this is the practice adopted by some of the best managed Internal Boards.

The pumps installed since 1919 show an increase in horse-power and capacity, not entirely accounted for by the necessary increase in lift. The average lift in 1913 was between 7 and 15 ft. Higher lifts are more generally met with to-day, rising to 21 ft. for the plants in the Swaffham, Bottisham, Littleport, and Downham districts, and even up to 26 ft. in the North Side district, Wisbech. This increased lift is only partly accounted for by the lowering surface; it is also due to the modern practice of deepening and enlarging the drains so that the water can be kept lower. This provides greater storage capacity, and at the same time gives better drainage. Some of these leading drains are of considerable size, so that the pumps need not run so frequently.

It is difficult to generalise about the size of pumps used. About 1926, the size rose to 42 in. pumps with a capacity of 150 tons per minute, requiring about 250 horse-power; but more recently the size of pumps installed has fallen to 24 in. with a capacity of 70 tons per minute. This is a reasonable size; the smaller unit is more economical for normal use, while the possibility of duplicating the plant allows for additional safety.

An outline of pumping installation in the Littleport and Downham district, a well-managed Board, gives some idea of the constant struggle to maintain internal drainage. The district has a taxable area of 26,000 acres, but the total area drained is of the order of 35,000 acres. It includes approximately 26 miles of drains. The district was among the first to adopt steam-driven pumps. A 30 h.p. engine and scoop-wheel was originally erected on the Ten-Mile River bank; this was increased to 80 h.p. in 1843. A similar engine had been installed on the Hundred Foot River bank in 1829, seven miles away. Both these were condensing beam-engines, the steam pressure being 15 lb. per sq. in., and the scoop-wheels about 41 ft. in diameter. Owing to the lowering surface, these scoop-wheels were increased in size to 50 ft. in 1882, and they weighed 75 tons each. In 1912, at the Ten Mile station, the Commissioners installed two double-acting open compound condensing engines of 200 h.p. directly coupled to 48 in. Allen pumps, each with a capacity of 150 tons per minute. While in 1914, at the Hundred Foot River station, a 400 h.p. steam-engine was installed to drive a 50 in. Gwynne pump handling 212 tons per minute against a total head of 21 ft. This was supplemented in 1925 by a Mirrlees Diesel engine with a 36 in. Gwynne pump lifting 110 tons per minute. Finally, in 1937, the Ten Mile set was further improved; 340 h.p. Allen engines replaced the older unit, but the existing pump casing was maintained.

Irregular surface lowering may cause an entire change in the direction of the drainage. That has been the fate of the Methwold and Feltwell Board. To aid the original drainage through Sam's Cut, a pump was installed in 1883 where the cut joined the Ten Mile River at Hunt's Sluice.[1] In 1913, it was felt advisable to erect an additional pump. This was the first crude oil engine in the Fenland, and was installed by the Campbell Gas Engine Co. By 1928, however, this ancient system had to be abandoned, and the drainage was taken across country to the River Wissey. A Mirrlees engine and an Axial flow pump (another newcomer) was installed, and this has now been supplemented in 1938 by two Allen engines and pumps, developing, between them, a total of 260 h.p. and capable of pumping 170 tons per minute. This improved drainage has necessitated a deepening and widening of the drains, which now flow in an opposite direction to the original layout. Certain drains and culverts have had to be abandoned because they can no longer function owing to the lowering of the peat surface.

[1] See p. 189 above.

## ADMINISTRATION

The outstanding difficulty of the past has been the lack of a single controlling authority and the absence of co-operation amongst existing authorities. The problem of finance has always been a very great stumbling-block to concerted action. In 1850, there were three principal bodies controlling the southern Fenland: (i) The Bedford Level Corporation controlling the North Level; the Hundred Foot and Old Bedford Rivers; and the Middle Level. (ii) The South Level Commissioners controlling the South Level as then defined. (iii) The Eau Brink Commissioners controlling the remaining portions of the Tidal River (below Denver), but with rights vested in the several *ad hoc* authorities.

In 1858, the North Level separated and now falls principally within the purview of the Nene Catchment Board. In 1862, the Middle Level separated from the Bedford Level Corporation. Bills were introduced into Parliament in 1877, 1878, 1879, and 1881, concerned with the idea of setting up Conservancy Boards, but difficulties of rating were among the main reasons that prevented their establishment.

The functions of the Eau Brink Commissioners were, in general, divided between the Ouse Banks Commissioners, the Lower Ouse Drainage Board, the Ouse Outfall Board, the Denver Sluice Commissioners, and the South Level Commissioners. In addition, there existed the Norfolk Estuary Company, which built and controlled the Marsh Cut and training walls; and also there was the King's Lynn Conservancy Board, mainly concerned with the port of King's Lynn and its navigation. In 1920, the functions of the Bedford Level, South Level, Denver Sluice Commissioners, Ouse Outfall Board, Lower Ouse Drainage Board, and the Ouse Banks Commissioners, were transferred to one Authority termed the Ouse Drainage Board, which had powers of direct rating over the area controlled.

Under the Land Drainage Act of 1930, the River Great Ouse Catchment Board was instituted to take over the functions of the Ouse Drainage Board. It also had control over the remaining portions of the River Ouse Catchment Area (excluding the Norfolk Estuary Company and the King's Lynn Conservancy Board), but without the powers of direct rating. The area covered by the Board is over two million acres, having a rateable value of over £3,350,000, and running into twelve administrative counties. Its income is derived partly from precepts laid on the Internal Boards as far as may be considered fair, and partly from precepts laid on the County Councils, with a statutory limit of 2*d*. in the £. It has been found necessary to levy this full rate, as the total income from the two sources is not sufficient to meet the onerous duties which fall on the Board.

The area covered by the Internal Drainage Districts is approximately 276,063 acres, with an annual value, for drainage rates, of about £635,000. These Internal Drainage Boards, which number 90, are responsible for local drainage.

The total mileage of main river for which the Board is responsible amounts to about 500 miles, of which 306 are in the Upland Area, and 190 in the Fenland itself. Work in the Upland Area rivers is similar to that carried out by other Catchment Boards, but with the strict necessity of bearing in mind the fact that these rivers discharge into the Fenland. In upper Cambridgeshire, the Board is responsible for the Rivers Cam and Rhee, which are now kept in as satisfactory a state as the funds will permit. The fenland areas are roughly at Ordnance Datum level, and the upland streams are carried through this area as embanked rivers above the level of the land on either side. The Internal Drainage Boards have, therefore, to pump the water up into the river above their own ground, and the pumps must be capable of lifting to a height sufficient to reach the flood levels within the banks.

## THE MIDDLE LEVEL

The Middle Level lies between the Old Bedford River and the River Nene (see Fig. 47). It contains 165,000 acres of ground, of which about 120,000 acres are actual fen with 150 miles of main waterways. Most of its fifty Internal Districts discharge by pumping into the Middle Level Drainage System, for which the Middle Level Commissioners are responsible. But the area of the Sutton, Mepal, Manea, and Welney Internal Boards has separate pumping stations discharging into a Counterwash drain that runs parallel with the Old Bedford, separated from it by a low bank. This drain has a gravity outfall just below Denver Sluice.

Some portions of the Middle Level fall to 4 ft. *below* Ordnance Datum; and it must be remembered that the high-tide levels outside rise to something of the order of 13 or 17 and even to 18 ft. above O.D. Prior to 1848, the Level discharged by Tong's Drain into the Ouse below Denver; but, on the advice of Messrs Burgess and Walker, a twelve-mile cut was made through Norfolk Marshland, and the Middle Level waters discharged into the Ouse through a new sluice at St Germans, eight miles below Denver Sluice. The cost of this scheme amounted to £450,000.

The reduced low-tide levels that resulted from these measures proved satisfactory for many years. In 1862, however, the sluices at St Germans "blew up"; the river banks broke down, and tides flowed up the cut to flood some 6000 acres. Sir John Hawkshaw devised a dam with a series of

16 syphons, each 3 ft. in diameter. This arrangement lasted until 1880, when a new sluice was erected to take the place of the syphons; the total cost of this disaster, and the improvement of 1880, amounted to £250,000.

By 1912, low water in the Ouse was not as low as it had been in 1880, due to a general deterioration of the river and its outfall into the Wash. In the meantime, the old scoop-wheels had been replaced by modern pumps so that the water had to be got away more quickly. The floods of 1916, 1923, and 1926 confirmed the Middle Level Board in its opinion that the position in time of heavy flood was becoming more dangerous. In 1923, Major R. G. Clark, as Engineer to the Commissioners, recommended the installation of improved sluices; but by 1928 it was decided to install a pumping station at St Germans, and this, after due negotiation, was completed in 1934. The new sluice has only two sluice gates each 35 ft. in width, thus securing 50 per cent greater width for discharge than in 1880. These sluices are assisted by three pumping units, and provision has been made for the installation of a fourth. The pumping plant was erected by the Premier Gas Engine Co.; each unit consists of a horizontal eight-cylinder Diesel engine developing 1000 h.p. driving a Gwynne centrifugal pump 8 ft. 6 in. in diameter, capable of discharging up to 1000 tons per minute at low heads or 840 tons per minute against a static head of 10 ft.

## THE SOUTH LEVEL

The problems of the Middle Level, intricate as they are, are very much simpler than those of the South Level. Those of the Middle Level are concerned mainly with the rainfall that falls on its own area, while the South Level has to arrange for the drainage of nearly one-half of the catchment basin of the Ouse.

Before the institution of the Ouse Drainage Board, the maintenance of banks was the responsibility of the Internal District Commissioners. During high floods, the better drained and richer areas were fairly well protected, because they had been able to maintain their banks in a satisfactory condition; but Internal Districts, whose financial position was not so strong, were liable to breaches in the banks. Some districts were flooded at frequent intervals; Hockwold Fen, for example, was drowned in 1912, 1915, and again in 1916. In 1919, three breaches occurred in the River Cam, and in 1928 there was a serious breach in the right bank of the River Wissey, when some 2000 acres were flooded. It was about this date that the Ouse Drainage Board received a grant of £276,000 from the Ministry of Agriculture to enable it to carry out extensive dredging and embanking throughout the South Level.

By 1934, this sum had been expended, and it became necessary to prepare a supplementary scheme of £103,000 to carry on the work. Early work under both these schemes was mainly confined to dredging. The dredged material was not of much use for embankment work, and for this purpose the Board, following the practice of the Internal Boards, obtained its clay from the Roswell Pits near Ely. Photographs of the pits in 1913 show work being done by hand, but gradually operations have been mechanised and fully organised. It has also been found economic to open up subsidiary pits. As a general rule, the banks are heightened and breasted or faced with clay to prevent them being washed away under wave action caused by high winds on the flood waters.

In floods prior to that of 1937, one bank at least had always broken. But during the floods of March 1937 no breach of any consequence took place, so that water levels in the streams rose higher than hitherto. The danger is that these high-water levels create a head sufficient to force water *under* the banks. It is therefore felt advisable to strengthen the banks, and future work will carry the clay breasting down the front of a bank by trenching on to the clay below. Previously, in weak places, this has been done by hand; but now, with a new and more extensive programme, it is being undertaken by trenching machines.

The floods of 1936 and 1937 yielded much valuable data, from which it has been possible to re-design the section for the main river from Littleport to Denver, to which all the other rivers in the South Level are tributary. This stretch is to be widened and some half a million yards will be dredged away. It is calculated that this widening will reduce flood conditions at Littleport, when the river is discharging, by a matter of 10 in. This in turn will provide greater storage capacity for the periods when Denver Sluice is closed by tidal waters.[1] The cost of the scheme will amount to £266,000, and a 75 per cent grant has been obtained from the Ministry.

The Hundred Foot River and the Old Bedford River are also receiving attention. The Middle Level Barrier Bank, which protects the Middle Level area from the flood waters from the Uplands, was heightened under a scheme completed in 1933. The Old Bedford River, too, is now being improved; and, consequent upon damage during the unprecedented flood of 1937, most of the Middle Level Barrier Bank is being protected with clay at the cost of about £60,000.

Denver Sluice was partially remodelled in 1923, when one large eye, 34 ft. in width, was installed instead of two smaller discharging sluices.

[1] See p. 193 above.

The result is that Denver Sluice can now take the full discharge from the South Level without any loss of head.

The outlet from the Washlands between the two Bedford Rivers is by means of Welmore Lake Sluice.[1] This was rebuilt in 1930, and subsequent observations, taken during flood conditions, have shown the advisability of increasing by 50 per cent the discharge capacity at this point, by the installation of a third sluicegate 24 ft. wide, which it is hoped will be constructed next year.

## THE TIDAL RIVER SECTION

The so-called tidal river section (i.e. the estuary below Denver Sluice) has been the subject of much controversy and of many reports during past centuries. The Eau Brink Cut of 1821 and the Marsh Cut of 1852 were especially successful because they shortened the length of the river.[2] Much benefit also resulted from the activities of the Norfolk Estuary Co., which was compelled by an Act of Parliament to carry training walls through Vinegar Middle Shoal (in the estuary) before it commenced reclamation. The walls were completed in 1857, by which time the company had spent £250,000 on the improvement of the estuary. The Norfolk Estuary Co. was intended originally to recover land from the Wash. Fig. 53 shows the result of its activity and of similar effort in Lincolnshire during the nineteenth century.

After a series of flood years during the nineteenth century, Mr W. H. Wheeler was consulted by the Denver Sluice Commissioners; and his report, issued in 1883, recommended that the river should be widened from Denver Sluice down as far as the Eau Brink Cut. The Eau Brink and Marsh Cuts were, apparently, in very good condition at that time. There were comparatively low-water levels under normal conditions, but the river was not wide enough to deal with flood waters—hence the necessity for the report. Had Mr Wheeler's scheme been carried out at the time, the result should have been very satisfactory, but unfortunately the banks and channels of the Wash were changing, and, in consequence, the estuary conditions have become steadily worse.

Comparison between Fig. 54 and Fig. 55 will show the change in the channels of the Wash between 1871 and 1936. The tide on the eastern side of the Wash follows a circulatory movement in an anti-clockwise direction and the channels follow this tidal flow. Thus, the water flowed in by the Lynn Channel and, following this circulatory motion, discharged by the Bulldog Channel. The channels were then well defined. At some later

[1] See footnote 4, p. 182 above. [2] See p. 191 above.

period the Teetotal Channel widened and the Daseley Channel broke through, while at the same time, for some cause which is as yet unknown, the inward end of the Bulldog Channel began to silt up. The full implication of these changes is not yet known; further investigation is now proceeding.

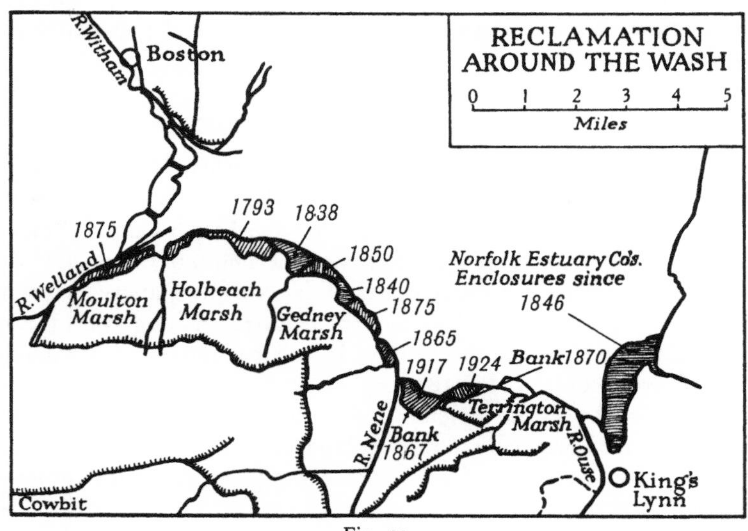


Fig. 53.

The coastal "marshes" were recovered before the nineteenth century. The inner banks, marked by toothed lines, represent the limits of still earlier enclosures.

Further flooding of the Fenland occurred during the period of the Great War, and, at the request of the Lower Ouse Drainage Board, Mr Havelock Case issued a report in 1917. Mr Case, like Mr Wheeler, recommended the widening of the river from Denver to the sea; he also advocated the installation of a larger sluice at Denver, which, as noted above, was not carried out until 1923. He also found that the conditions in the Wash were so bad as to require the further construction of training walls.

The position was again discussed in 1918, when Mr Preston held an enquiry at King's Lynn. A proposal was put forward for the construction of a barrage instead of training walls, but the enquiry showed that there was not sufficient technical evidence to make a decision.

In 1925, a Commission of enquiry was set up by the Ministry of Agriculture, and their technical adviser, Mr Binnie, put forward a scheme for training walls. He believed that the training walls should be sufficiently high to carry the river water through to the Hull Sand Beacon, some

5 miles out to sea. His contention was that the silting of the river was due to material in suspension and that this should be excluded. A further report was subsequently prepared by Sir Alexander Gibb, but the financial burden was impossible for the Ouse Drainage Board to carry.

By 1930, the original short training walls had seriously deteriorated. But the Ouse Drainage Board felt that it could not finance any improvement, which was, therefore, undertaken by the Lynn Conservancy Board with Government assistance. The principle adopted was to protect the toe of the existing walls by means of brushwood mattresses and to heighten the tops by a combination of brushwood and stone. This particularly suitable method of repair was put forward by a firm of contractors of Dutch origin, who were operating in this country. Subsequently, the firm called in two eminent Dutch engineers, and the result was the so-called "Dutch Scheme", which was considered by the Labour Government in 1931. By this time, conditions in the estuary had become so bad that it was estimated that it would cost five and a half million pounds to put it right.

The "Dutch Scheme" provided, as before, for the widening of the river from King's Lynn to Denver and Welmore Lake Sluice. It also advocated the cutting through of Magdalen Bend and the widening of the Hundred Foot River. Because the upper section of the tidal river had probably been silted up by material coming in from the Wash, it was proposed to construct a set of sluices across the Hundred Foot River in the neighbourhood of Welmore Lake Sluice. From here down to the sea the toe of the banks was to be protected by mattresses, and their slopes pitched with concrete blocks. To take the river out to deep water, training walls over 5 miles long were suggested. The lower portion was to be made with mattress work, and the upper portion was to consist of caissons of concrete blocks. The height was to be brought up to at least neap-tide level, and, in order to regulate the depth of channel, a series of groynes was to be constructed inside the new walls. Although the Government offered a 90 per cent grant, it was again felt that the Catchment Board which had just come into existence, in 1930, could not face the financial burden.

But, at this time, the Catchment Board had to face the reconstruction of training walls for a length of 1 mile on the eastern side. This had to be done under difficult financial conditions. Work was undertaken in 1932 and cost the Board £85,000. It was felt advisable to reconstruct a new wall slightly behind the old wall, and to carry it up to a somewhat higher level.[1]

During the dry summers of 1934–35, silt travelled[2] steadily up-river

[1] Both the east and the west training walls were extended a short distance in 1937.

[2] See p. 189 above.

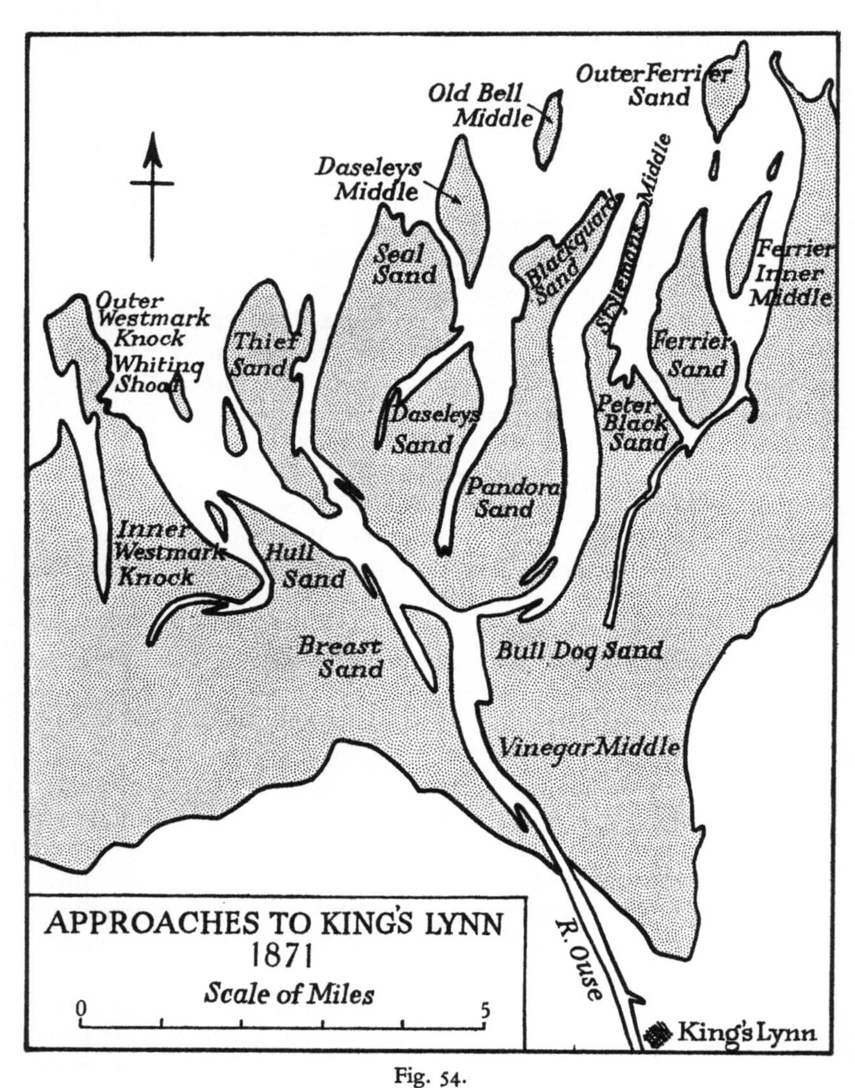


Fig. 54.

Based upon British Admiralty Charts.

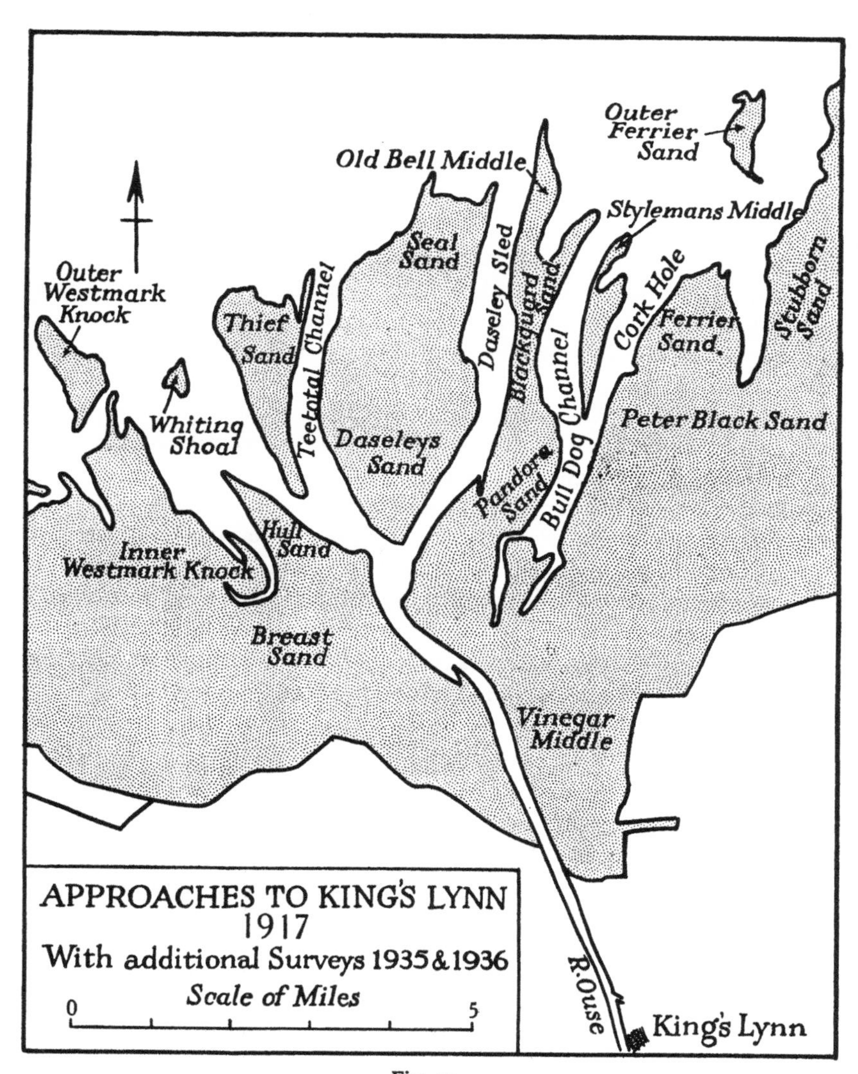


Fig. 55.

The 1917 data were based upon British Admiralty Charts. The data of 1935 and 1936 were obtained from the surveys of the River Great Ouse Catchment Board.

until the bed at Denver had risen some 8 ft.; indeed, at certain inlets the bed had risen a matter of 13 ft. This meant a blocking-up of the river by over a million cubic yards. The flood conditions of 1936 scoured out a certain amount of this accumulation, otherwise the flood of 1937 would have been disastrous. The general effect of this heavy flood of 1937 was to scour down the river bed for a distance of about 8 miles below Denver. A great quantity of silt must have been carried out to sea, although the percentage of silt in suspension in the river current in the Wash seems to have been somewhat low. A vast quantity of silt, however, did not reach the sea, but was deposited in Marsh Cut, which, due to neglect, has been steadily widening since its original construction. In fact, the section of the Marsh Cut is now about 60 ft. wider than it was in 1860, and its bed has risen some 8 to 10 ft.

The bed of Marsh Cut is now higher than the bed of the river upstream; and, in order to remedy this state of affairs, the Board (with the assistance of the Ministry) has agreed to a scheme (1) for lining the banks with stone pitching to prevent further erosion, and (2) for the construction of groynes throughout the length of the Cut to provide a narrowing channel at low-water level. This work, extending for about 4 miles to the Free Bridge at King's Lynn, is being done by contract at a cost of one-quarter million pounds. Although, as the result of deposition, the bed between the training walls out to sea rose a certain amount, it is still lower than that of the Marsh Cut, thus showing the benefit of the training walls.

## THE TIDAL MODEL

Since 1932, the Catchment Board has been engaged on an investigation of the many problems of the estuary. For this purpose, a large tidal model has been built at Cambridge; this, at the time of its construction, was the largest tidal model in the world. It is essentially a model of the Wash out to a line drawn from Hunstanton to Friskney Flat near Skegness.

The technical particulars[1] are as follows:

| | |
|---|---|
| Horizontal scale ... | 1 : 2500 |
| Vertical scale ... ... | 1 : 60 |
| Vertical exaggeration ... | 1 : 41·7 |
| Time scale ... ... | 1 : 324 |
| Velocity scale ... ... | 1 : 7·7 |
| Tidal period ... ... | 138 seconds |
| One year of tides ... | 27·1 hours |

The tides are produced by a plunger, weighing 14 tons, which displaces water from a large trough; the water, flowing over the model area of the

[1] These particulars are taken from the brochure issued in connection with the Model.

Wash, correctly reproduces the tide. The mechanism operating the plunger incorporates a special cam which enables the production of the correct tidal cycle to be obtained. The flow from the various rivers is controlled by valves which may be set, so that a correct volume of water flows down the model channels which accurately represent the actual rivers.

The correlation of the model hydraulic conditions with those found in nature was undertaken first from data collected by the Admiralty in 1917, and again from data collected by the Board in 1935. The model was moulded to reproduce these conditions respectively. The satisfactory results of these correlation tests enabled those in charge of investigations to take a step forward, and to determine the changes in hydraulic conditions (and in the configuration of the channels) likely to result from schemes carried out in the rivers of the Wash. Tests of certain proposals have been made.

Results have shown that the problem is much more involved than appears at first sight. A careful study is being made not only of the conditions in the Wash, but of similar works which have been undertaken elsewhere, particularly in Holland and Germany. Investigation has been somewhat handicapped by the lack of finance. For the design of the groynes in the Marsh Cut, together with other alterations proposed in the section running through to Denver, a secondary model is under construction. This will have a horizontal scale of 1 : 240 and a vertical scale of 1 : 100. As the model includes the length of river up to Denver, it will be about 360 ft. long, and must be constructed in the open air. The plunger for the creation of the tide weighs 4½ tons; and the tidal period will be 31 minutes. This secondary model will enable experimental work to be done on a larger scale.

CHAPTER FIFTEEN

# THE BRECKLAND

## By R. R. Clarke, J. Macdonald, and A. S. Watt

## (A) HISTORICAL AND ECONOMIC BACKGROUND

## By R. R. Clarke, B.A.

BRECKLAND IS A NATURAL REGION UNIQUE IN BRITAIN BUT paralleled in Western Europe by the heaths of Denmark, Holland, north-west Germany, and the Rhine Valley.[1] Roughly speaking, it covers some 400 square miles in the counties of Norfolk and Suffolk, while its south-western extremity impinges on the eastern boundary of Cambridgeshire.[2] The precise limits of the region are not easy to define, for, save on the south-west where it marches abruptly with the peat and clay of the Fens, the Breck district shades imperceptibly into the regions of gravel and chalk that elsewhere surround it. The borders of Breckland therefore present mixed physical characters with many outliers (Fig. 57), but Fig. 56 will serve to indicate the location of the main area.

This main area is mostly a low plateau rising between 100 and 200 ft. above sea level. It owes its geographical personality to a remarkable pall of sand that covers its complex sub-soil of chalk, gravel, sand, loam and chalky boulder clay out of which the lime has been dissolved by rainwater. The Breck soil is, with a few insignificant exceptions, arid and highly permeable. Combined with a relatively dry climate,[3] these characteristics have produced its peculiar vegetation and fauna, and have controlled human activity in the district. But for human agency, Breckland would be, under present climatic conditions, largely a treeless steppe, and any woodland that might flourish would be open in character and free from scrub. Besides its characteristic vegetation, and insects,[4] the Breckland heaths are important ornithologically,[5] and form one of the chief strongholds of the stone curlew and the ringed plover, while other rare birds nest by its heathland pools and meres.

In this arid region, human settlement is very dependent upon water

[1] Full bibliographies of recent work on the district are contained in: (1) W. G. Clarke, *In Breckland Wilds*, second revised edition by R. R. Clarke (1937); (2) H. Schober, *Das Breckland: eine Charakterlandschaft Ost-Englands* (Breslau, 1937).

[2] Naturally, any boundary-line must be arbitrary. Country with some "breck" characteristics can be found outside the Breckland proper.

[3] See p. 43 above. [4] See p. 70 above. [5] See p. 62 above.

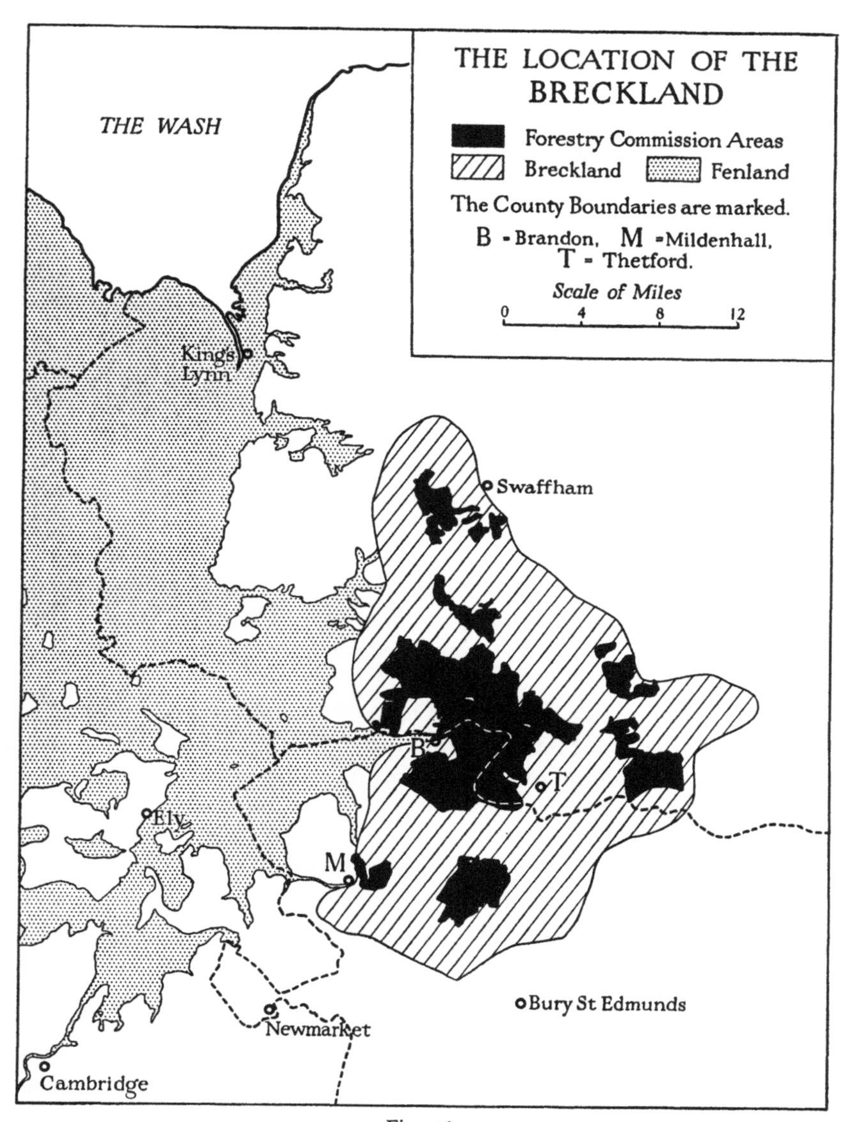


Fig. 56.

This figure shows the main location of the Breckland. Outliers with some typical Breck features occur beyond this arbitrary frontier, e.g. in Cambridgeshire, to the north of Newmarket.

supply provided by the valleys of the Wissey and the Little Ouse-Thet that run through the middle of the district, and by those of the Nar and the Lark, near its northern and southern margins—all of which empty into the Fens and so to the Wash. From prehistoric times, settlement has been focused on these valleys and their tributaries, and only four parishes (Swaffham, Elveden, Ingham, and Wordwell) appear never, in historic times, to have had access to stream or fen or mere. The meres of Breckland provide a tolerable substitute in the absence of rivers. The biggest and most typical of these curious sheets of water lie in five parishes in Norfolk, and the best known are Fowlmere, Langmere, Ringmere, and the Devil's Punchbowl; the largest of all, Mickle Mere (29¼ acres), is near-by in West Wretham Park. With one exception, the water level in these meres has no visible inlet or outlet, and is subject to remarkable fluctuations. At times, the meres are completely dry for several years; at other times, they overflow adjacent roads. There can be little doubt that their waters are derived from the surrounding chalk, and that they rise and fall with the saturation level in the underlying rock. Rainfall is thus solely responsible for their fluctuating levels. Some at least of the meres may have been formed from "pipes" in the chalk filled with drift-sand. The importance of the meres as sources of water is shown by the numerous parish boundaries which meet at them. At Rymer Point, 4 miles south of Thetford, no less than nine parishes meet, and here, formerly, was a considerable natural sheet of water.

The palaeolithic flint implements found in its gravels and brickearths; the important neolithic flint mines at Grime's Graves; the flint implements scattered by the million over the surface of its heaths and arable fields;[1] its extensive mileage of primitive trackways; its impressive dykes and its numerous barrows and other relics of early cultures which are constantly being discovered—all these evidences indicate that in some of the prehistoric periods Breckland must have been one of the most thickly populated districts in Britain.[2] The principal attraction of the region to early man lay in the absence of heavy woods which he was unable to clear. In addition, the margins of the Fenland and of the heathland meres yielded fish and fowl; while, for tool-making, the chalk provided unlimited quantities of the finest flint in Britain. Then, too, the Icknield Way,[3] along the chalk

[1] The working of flint in this district has probably been continuous from prehistoric times. To-day, Brandon supports the last surviving flint-knapping industry in Britain. The mines at Lingheath still produce some of the raw material required for the manufacture of gun-flints—see R. R. Clarke, "The Flint-Knapping Industry at Brandon", *Antiquity*, ix, 38 (1935).

[2] Well shown by the distribution maps of Sir Cyril Fox. See (1) *The Archaeology of the Cambridge Region* (1923), (2) *Proc. Prehist. Soc. E. Anglia*, vii, 149 (1933).

[3] See p. 85 above.

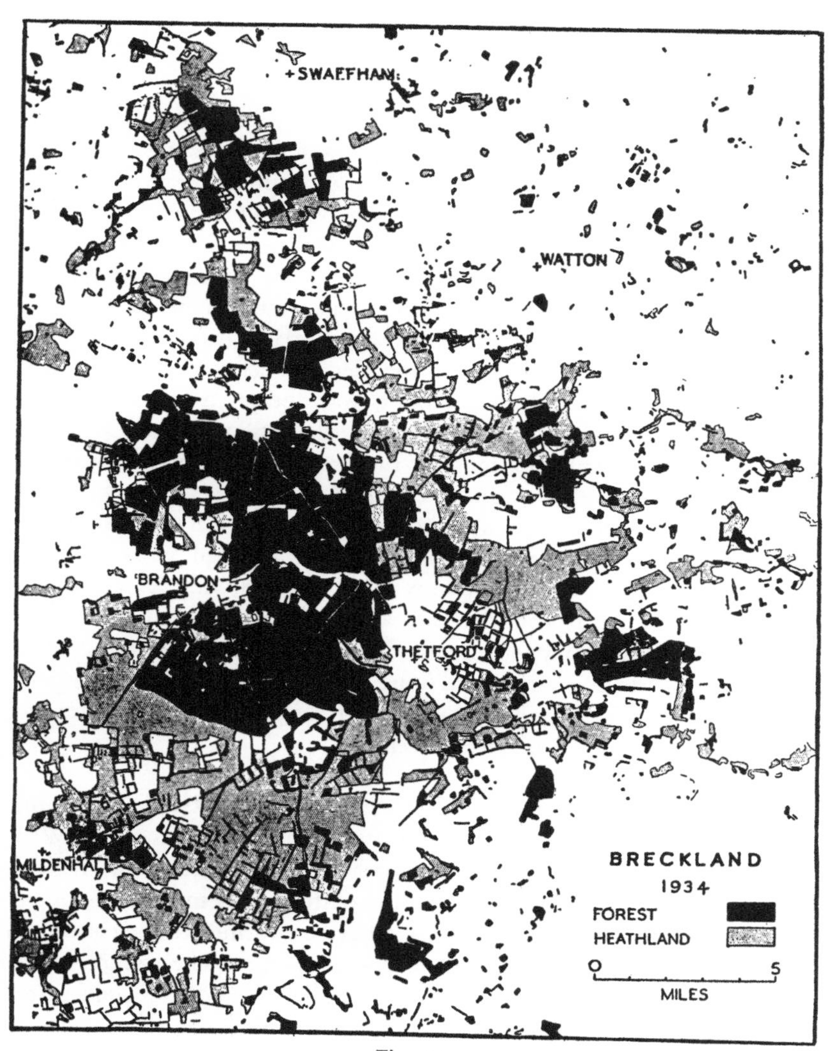


Fig. 57.

Reproduced by the courtesy of the Land Utilisation Survey of Britain.

ridge, offered easy intercourse with the rich cultural province of Wessex. It was probably during the Bronze Age that Breckland became one of the chief centres of population in Eastern England, but as Iron Age man acquired the power to subdue and exploit the more stiff but richer soils of adjacent regions, the cultural focus of "East Anglia" moved south-west leaving Breckland as a backwater for a thousand years. Not until the late-Saxon period did Breckland acquire a new strategic status, when the deforestation of the claylands of Norfolk and Suffolk again swung the economic pendulum north-eastward. Though still a poverty-stricken steppe, as the Domesday Book attests,[1] it was now the gatehouse of a wealthy East Anglia commanding the Icknield Way, still the main line of approach from the civilised south. Although a waste-land it was a frontier zone through which communication was essential. The rise of Thetford[2] to the zenith of its importance as the eleventh-century capital of East Anglia, with its cathedral and its mint, was due to its location on this highway, at the confluence of the Rivers Little Ouse and Thet.

Place-names indicate that most of the present primary settlements of Breckland are of Anglian origin; there are 8 *-ings* and *-inghams*, 20 *-hams*, 13 *-tons*, and 8 *-fords*. The importance of the rivers for water supply is demonstrated by the concentration of these nucleated villages in the valleys. Two villages are associated with the Nar, 28 with the Wissey, 16 with the Little Ouse, 10 with the Thet, and 9 with the Lark, each including its tributaries. Secondary settlements consisting of heathland farms, with their satellite cottages, and isolated houses for warreners and gamekeepers, only came into existence, in most cases, with the growth of enclosures and tree-planting during the nineteenth century.

What were the main features of the economy of Breckland prior to the modern enclosures? Recent investigation has shown that in West Wretham,[3] and also several other parishes in the heart of the region, something akin to the Scottish infield-outfield system was common, though the border parishes are likely to have conformed to the custom of the normal Norfolk and Suffolk village-community. The essence of the system was a division of the arable land of a village into two unequal parts: a small infield probably cropped continuously, near the village; and a larger outfield comprising five to ten temporary enclosures from the waste

[1] See H. C. Darby, "The Domesday Geography of Norfolk and Suffolk", *Geog. Jour.* lxxxv, 432 (1935).

[2] It is interesting to note that the Domesday Book numbers the burgesses of Thetford as 720, compared with 665 at Norwich and 70 at Yarmouth.

[3] J. Saltmarsh and H. C. Darby, "The Infield-Outfield System on a Norfolk Manor", *Economic History*, iii, 30 (1935).

(called brakes, folds or faughs), of which one was broken up every year, cropped continuously for a few seasons (with the aid of sheep manure and marling), and then allowed to revert to its former condition until its turn came to be ploughed again. Fig. 58 shows the fields at West Wretham in the mid-eighteenth century, and illustrates conditions generally. These outfields were large, but few can have equalled those at Northwold in the seventeenth century when men ploughed straight for 12 furlongs.[1]

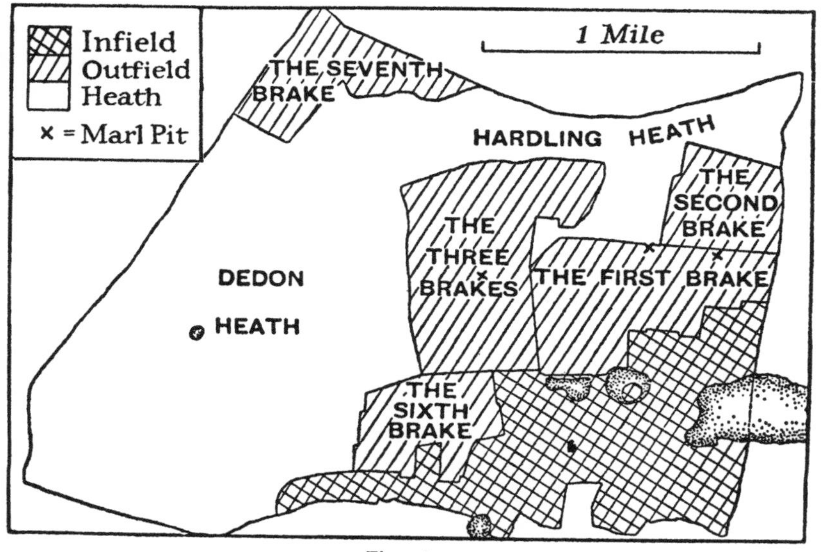


Fig. 58.

Field System at West Wretham (Norfolk), 1741.

From J. Saltmarsh and H. C. Darby, "The Infield-Outfield System on a Norfolk Manor", *Economic History*, iii, 34 (1935). This is diagrammatised from the original map on two sheets of vellum pasted together, and measuring $53\frac{1}{2} \times 36\frac{1}{4}$ in. It is preserved in the Muniment Room of King's College, Cambridge.

According to W. G. Clarke, "parts of almost every area of heathland were at one time cultivated, but have become derelict. Both these areas and the large sandy open fields are known as 'brecks', and their number, and the fact that they are characteristic of all parishes, induced me in 1894 to give the district the name of Breckland."[2] Thus it seems that the name by which the whole area is known may mean nothing other than "the land of outfields".

[1] Mentioned by Sir Philip Shippon, 1671, *Norfolk Archaeology*, xxii, 176 (1925).

[2] W. G. Clarke, *In Breckland Wilds* (1925), p. 22. The term "The Brock District" was used by Prof. A. Newton in the introduction to H. Stevenson's, *The Birds of Norfolk*, vol. i (1866).

The crops of the district were as characteristic as its field system and its waste lands. Rye was the commonest cereal, but the yield of barley was often the largest, with oats next. The wheat crop was small. Large flocks of sheep were kept in every parish for fertilising the soil while there was "no where better Mutton than this barren Land affords, the Sheep being not liable to the Disease called the Rot".[1] Pre-enclosure travellers were very impressed by the abundance of rabbits. "A large portion of this arid country is full of rabbits, of which the numbers astonished me", wrote the Duc de la Rochefoucauld in 1784. "We saw whole troops of them in broad daylight; they were not alarmed by noise and we could almost touch some of them with our whips. I enquired of this prodigious number and was told that there was an immense warren which brought in 200 guineas a year to the owner, being let to a farmer."[2] The penalties for poaching were severe, because the farming of rabbits formed the economic mainstay of many of the landowners. Some farmers still pay their rent from what they realise by the sale of rabbits, but the number caught is rapidly decreasing with the spread of afforestation, which necessitates the extermination of all rabbits within its confines.

One of the common features of Breckland in the pre-enclosure period was the prevalence of disastrous sandstorms. A notable storm in 1668 blew sand for 5 miles from Lakenheath Warren to Santon Downham, almost overwhelming the village and obstructing the navigation of the Little Ouse.[3] John Evelyn, in 1677, also referred to "the Travelling Sands about ten miles wide of Euston, that have so damaged the country, rolling from place to place, and, like the lands in the Deserts of Lybia, quite overwhelmed some gentlemen's whole estates".[4] The open and unrestricted appearance of this region, before enclosure and afforestation wrought such drastic changes in its scenery and economy, is well described by an eighteenth-century traveller, William Gilpin. Between Brandon and Mildenhall, he declared that:

> Nothing was to be seen on either side but sand and scattered gravel without the least vegetation; a mere African desert. In some places this sandy waste occupied the whole scope of the eye; in other places, at a distance we could see a skirting of green with a few straggling bushes which, being surrounded by sand, appear'd like a stretch of low land shooting into the sea. The whole country indeed had the appearance of a beaten sea-coast, but without the beauties which adorn that species

[1] F. Blomefield, *An Essay towards a Topographical History of the County of Norfolk*, i, 553 (1739).

[2] F. de la Rochefoucauld, *A Frenchman in England, 1784* (1933), p. 212.

[3] T. Wright, "A curious and exact relation of a Sand-floud, which hath lately overwhelmed a great tract of land in the County of Suffolk", *Philosophical Transactions*, No. 37 (July 1668).

[4] John Evelyn, *Diary*, 10 Sept. 1677.

of landscape. In many places we saw the sand even driven into ridges; and the road totally covered, which indeed was everywhere so deep and heavy, that four horses which we were obliged to take could scarce in the slowest pace drag us through it. It was a little surprising to find such a piece of absolute desert almost in the heart of England.[1]

It must be remembered, however, that casual travellers through the district may have exaggerated its wild and barren character, for the main trackways crossed the heathlands remote from the more fertile valleys.

Even so, if this barren soil was ever to be cultivated it was essential to plant trees. The enclosure movement, towards the close of the eighteenth century, was accompanied by the planting of belts of dwarfed hedges of conifers, especially of Scots pine, to shelter the fields from winds. But tree-planting on a large scale only began about 1840. The incidence of enclosure in Breckland varied with the soil, and its effects were more marked in the border parishes. There, holdings were consolidated into large estates, corn production was increased by more intensive cultivation, and population expanded rapidly. Sheep manure and marl had helped to feed the hungry sands of Breckland; the outfield rotation was a device for concentrating upon a small area the "tathe" of a flock supported by the grazing of the whole township.[2] But the introduction of the four-course shift of the new Norfolk husbandry brought changes. Under turnips and artificial grasses, the sand produced more fodder than ever before; more sheep could be carried to the acre; their "tathe" would consequently be richer and the crops heavier. It was probably the introduction of the new convertible husbandry that ousted the infield-outfield system from West Wretham.

But there were yet other changes to come. The agricultural crises of the nineteenth century from 1813 to 1837, from 1874 to 1884, and during the 1890's, saw the decline of arable farming, and the acquisition of vast estates by great landowners, a few of whom owned almost the whole of Breckland. One estate covered 34 square miles, another 20 and a third 18. Many tried to counteract their agricultural losses by developing the leasing of the sporting rights, and, to facilitate their disposal, tree-planting was encouraged as it provided cover for game. To-day, there is less land under the plough than there was one hundred and fifty years ago. This decline is due primarily to economic causes, but it may well have been hastened by soil impoverishment. Artificial manures on these poor soils are not always productive of good crops, while ploughing breaks up the chalk and assists its disappearance from the upper layers of soil. Fertility can then

[1] W. Gilpin, *Observations on several parts of Cambridge, Norfolk, Suffolk and Essex... made in 1769* (1805), p. 28.

[2] J. Saltmarsh and H. C. Darby, *art. cit.* p. 43.

be maintained only by marling or by introducing humus to absorb the artificial manures. Both mustard and lupins are often ploughed in for this purpose.

The tillage of poor land, like that of the Breck country, is lucrative only when prices are high, and this factor has encouraged experimental crops in what is, economically, a marginal area. In recent years, mature tobacco has been grown on the deeper sands at Croxton, Icklingham and Methwold, but the experiment failed, partly because the leaf could not be dried without artificial means. The introduction of sugar beet has been more successful as the sugar content is high, and beet is now the principal crop on soils which have been matured. Quite recently, black currants and asparagus for canning have been grown successfully on a large scale on the Kilverstone estate, where the light soils are fertilised with pig manure. Among the older established crops, barley is of most importance, though its yield is the lowest in East Anglia. Other crops are potatoes, lupins and mustard for sheep feed, buckwheat for game, lucerne and rape, peas, clovers, vetch and sainfoin, swedes, turnips and mangolds. Good pasture is rare even in the small fertile valleys, and so the density of live stock is only about half that of the adjacent districts. Cattle are few, though dairy animals have increased since the war, while sheep are below the average, being grazed usually on mere rough pasture.

Significant as are recent attempts to increase the agricultural and horticultural productivity of Breckland, they are less interesting than the post-war afforestation. This is the most fundamental vegetational change in the region in historic times, equalled only by the planting and enclosing of its treeless, grassy steppes at the close of the eighteenth and the dawn of the nineteenth centuries. To-day, the largest single forest area created in Britain in modern times is growing to maturity, and has wrought a revolution in the natural and economic equilibrium of the region.

Breckland is now the least densely populated region of its size between the Pennines and the New Forest. With the stimulus of enclosures, its population rose during the early nineteenth century, and, despite agricultural depression, this reached a total of over 40,000 in 1851. The subsequent depressions helped to depopulate the countryside, though the towns of Thetford, Brandon, Mildenhall, and Swaffham maintained the position they gained in the earlier part of the century. By 1931, the total population of the area was only just over 30,000; and, if the urban population of about 12,000 is subtracted, the remaining 18,000 are scattered over its heaths and valleys at about sixty to the square mile—less than one-tenth the average density for England and Wales.

## (B) AFFORESTATION IN THE BRECKLAND

By J. Macdonald, B.SC.

*Divisional Officer, H.M. Forestry Commission*

The Forestry Act of 1919 initiated something new in the rural economy of this country when it set in motion the work of afforesting large areas of land under the control of the state. The Forestry Commission, established under the Act, was set the task of safeguarding the national supplies of timber and other forest produce by creating in Great Britain an area of woodland large enough to tide the country over a period of emergency of about three years. For this purpose it was estimated that, in addition to the existing areas of woodland, mostly privately owned, it would be necessary to create about one and three-quarter million acres of entirely new forest.

The land for this enterprise must be capable of growing trees to a size at which they can be utilised. This means that large areas of land in this country, at present contributing little to the national resources, have had to be excluded from consideration because they are too exposed and high lying and because their soils are too poor even for the less exacting species of tree. On the other hand, it would not be in the national interest to include good agricultural land, although this is generally capable of growing excellent trees, particularly hardwoods. Consequently, the land for planting has been sought where possible in areas which are not too exposed or too high lying, and yet which are uncultivated or on the margin of economic cultivation.

The Breckland is a good example of the type of country into which forestry can be introduced without displacing, or threatening, any vital national interest. So far as can be known, the district has always been poorly wooded, although the plantations which have been made since the middle of the eighteenth century have shown that numerous species are capable of making good growth and reaching timber size. A great part of the area has not been cultivated within memory but has remained as open heath, formerly a pasturage for sheep but given over latterly to game and rabbits.[1] It is estimated that about 19 per cent of the total area planted by the Commissioners in Breckland has been at one time or another under the plough. Much of this, however, has in recent years been cropped only for game feed, while a considerable part was broken up during the war under the "Food Production" schemes.

[1] See p. 214 above.

Work began in the district in the winter of 1921–22 south of Brandon and at Cockley Cley near Swaffham. Since that time, the acquisition, and planting, of land have gone on steadily, and at the present time the Commissioners are in control of an area of 52,807 acres almost wholly in Breckland proper (see Fig. 56). There are three Forest units—Thetford, Swaffham, and the King's Forest; and details of these are given in the following table.

| Forest | Total area acquired | Total area planted to 30. iv. 1938 |
| --- | --- | --- |
| Thetford | 40,869 | 28,875 |
| Swaffham | 5,948 | 4,879 |
| The King's | 5,990 | 2,421 |

Thetford Forest is now the largest planted area in England. It is composed of a central block which extends from Methwold, on the north, to Elveden, and from Hockwold, on the west, to Croxton, together with outlying areas at Hockham, West Harling, and Mildenhall. Swaffham Forest consists of various blocks to the south-west and south of the town of Swaffham, as well as of an area at Didlington, north of Mundford. The King's Forest lies to the north of Bury St Edmunds. It was acquired in 1935 and owes its present name to its selection as one of the forests chosen to commemorate the Silver Jubilee of King George V.

In all three forests planting has been carried out mainly with coniferous species, among which Scots pine and Corsican pine preponderate. There are several reasons for this concentration on conifers. In the first place the Commissioners must pay attention to the type of timber that is most required in industry. At the present time, more than 90 per cent of the timber and wood products used in this country comes from softwood or coniferous trees, and there is no sign that in the future there will be any marked change in this proportion. It is reasonable therefore that most of the planting, not only in Breckland, but all over the country, should be done with conifers.

In the second place, the soils of Breckland are generally suitable for the growth of conifers. These soils, it is true, vary widely and form, roughly, a series running from a very thin sand over chalk, to deep, podsolised sands on some of the heaths where chalk is a long way from the surface.[1] On the last type, conifers are the only choice. On soils where chalk is at a moderate depth, and where sufficient soil moisture is available, good crops of oak could be raised; while on the thin soils, immediately over the chalk, beech would grow, although it is nowhere very vigorous in this district.

[1] See p. 223 below.

But, for reasons connected with the local climate, these and other broad-leaved species, as well as conifers such as Douglas fir and European larch, are often extremely difficult to establish when planted in the open. The two principal factors which work against them are frost and drought. Frosts in the late spring and early summer are a normal feature of the climate of Breckland. These are often quite sharp (up to 10 degrees of frost is not uncommon in the second half of May), and they fall with great severity on oak, beech, and other hardwoods. Douglas fir and European larch also suffer in the same way though less severely.

Drought also plays an important part in checking the growth of young broad-leaved trees and of conifers such as the larches. The rainfall is low and the soil unretentive of moisture, while, on the grass-covered areas, the dense sward leads to an intense local competition for supplies of moisture and, at the same time, acts as a covering which prevents much of the rainfall from actually reaching the soil. The high temperatures of the summer also tend to have an injurious effect on beech planted in the open, as this species is apt to suffer from sun scorch. The pines, and especially the Scots pine, are remarkably resistant to frost and drought, and for this reason alone they are likely to remain the principal species used in new planting in Breckland.

At the same time the importance of establishing broad-leaved trees on those soils which are really suited to them has not been overlooked and much experimental work is being carried out on this subject. In particular, methods of introducing beech into young plantations of pine have been studied in some detail.

The earlier plantations of Scots pine dating from 1922 and 1923 are developing rapidly, although there is much variation in growth, which can probably be correlated with variations in the soil. Already, there are trees up to 25 ft. in height, and a preliminary thinning has yielded produce in the form of fencing stakes, pit-props, small poles, and firewood. The pine plantations have not suffered much damage from fungi apart from the death of small groups of trees apparently killed by *Fomes annosus*, and by the fairy-ring fungus, *Paxillus*.[1] Damage by insects has however been more severe, and, in the Scots pine plantations, the pine shoot moth *Evetria buoliana* has been a dangerous pest for a number of years.[2] The attacks of this insect have led to the distortion of a large number of trees, and, although the damage has turned out to be less serious than was at one time feared, it has been sufficiently important to make special treatment of

[1] T. R. Peace, "Destructive fairy rings associated with *Paxillus giganteus* in young pine plantations", *Forestry*, x, 74 (1936).

[2] *Studies on the Pine Shoot Moth*, Forestry Commission Bulletin, No. 16 (1936).

the crops a necessity. The Corsican pine, although not immune to attack, is much less frequently damaged, and plantations of this species are full of vigorous, straight poles.

Most of the plants used in the afforestation in Breckland have been raised locally in the Commission's own nurseries, the most important of which are situated at Weeting, Lynford, Harling, and Santon Downham. These extend to 88 acres, and at the end of September 1937 they contained five million transplants and fourteen million seedlings. About eight thousand pounds of seed are sown annually.

One important side of the work is protection against fire, which, on account of the dry climate and the inflammable nature of the crop, is a serious menace particularly in the spring and summer. There are two observation towers manned in periods of danger, and connected by telephone with the central office at Santon Downham; while patrols are also put along the roads. In order to prevent fires spreading from road or railway, strips are ploughed and kept free from vegetation, while similar strips are also ploughed along rides through the forest. These ploughed strips are generally sufficient to prevent the spread of a ground fire in its early stages. Broad-leaved crops are less inflammable than conifers, and for some years it has been the policy to plant belts of hardwood trees along roadsides to serve as a protection. These will also have the effect of adding to the amenity of the countryside, and to promote this, various ornamental trees like the red oak and the wild cherry are now being planted, in addition to the common species such as oak, birch, beech, and sycamore.

In addition to their programme of afforestation, the Commissioners were charged with land settlement, which they have been carrying out by means of their forestry-workers' holding schemes. Holdings are generally created in the proportion of one for every 200 acres. Each holding consists of a house, buildings, and land which does not as a rule extend to more than ten acres. The holders are guaranteed 150 days' work in the year, but many of them obtain almost full-time employment. There are at present 188 of these holdings in the Breckland forests, and they house 630 persons, of whom 134 are workers in the plantations. At the end of 1937, the value of the live stock on these holdings was estimated at £7980. The number of workmen employed by the Commissioners in Breckland during 1937 varied between 300 in the winter months and 225 during the summer months, when the amount of work available normally falls off. In addition to the holdings there are within the boundaries of the forests twenty-four farms let on agricultural tenancies. These are not likely to be planted.

## (C) THE ECOLOGY OF BRECKLAND

### By A. S. Watt, PH.D.

From west to east in England, as the oceanic influence decreases, there is a fall in the Atlantic element of our flora, and a new element—not homogeneous, but commonly referred to as the "continental" element—becomes significant. Breckland is its headquarters in this country;[1] on the Continent it ranges from the far north of Europe to the south, and eastwards to the steppes of Russia and beyond. In a climate which is permissive to it, two other sets of factors condition its survival, namely, a soil with a high base status and/or freedom from competition. This last is freely offered by abandoned arable fields, disturbed soil, and open communities.

Freedom from competition very likely explains the presence of the liverworts, *Lophozia barbata* and *L. hatcheri*, and the lichens, *Cladina rangiferina* and *Stereocaulon evolutum*, in this outpost to the south-east of the main area of their occurrence in this country. It also explains the high percentage of annuals in the flora: half the "continental" element are annuals, and so are 40 per cent of the flora of the grasslands described later in this chapter.

The flora is essentially heliophilous and xerophytic. The annuals are drought evading, the perennials drought resistant. In soil preferences, there is a wide range represented, but calcicoles and species of slightly acid soils are numerous, while calcifuges are few and there are noteworthy absentees, e.g. *Erica cinerea*. The same numerical representation characterises both the continental element and the annuals: the bulk of each class is found on soils with a relatively high base status, a few only grow on very acid soils. But interest in the rarer species ought not to blind us to the fact that most of the species in Breckland have a wide English and British distribution.

### THE VEGETATION

While the interest of the flora of Breckland is enhanced by the presence and frequency of the continental contingent, the dominants of the vegetation do not suggest continentality: rather the reverse. These dominants are *Festuca ovina* and *Agrostis*[2] spp., the chief constituents of the variable

[1] See p. 43 above; and also A. S. Watt, "Studies in the Ecology of Breckland. I. Climate, Soil and Vegetation", *Jour. Ecol.* xxiv, 117 (1936).

[2] The species of *Agrostis* require revision as the result of W. R. Philipson's work. *Jour. Linn. Soc.* li, 73 (1937).

"grass-heath", *Carex arenaria, Calluna vulgaris, Pteridium aquilinum* and, locally, *Ulex europaeus.* Of these species, *Carex arenaria* and *Ulex europaeus* are West European; while *Calluna vulgaris,* although it stretches far eastwards to the plains of Russia, attains its best development in the west; and the cosmopolitan *Pteridium aquilinum* tends, in those parts of Europe with a continental climate, to become a woodland plant. But the behaviour of some of these plants shows an insecurity of tenure suggesting that as dominants they are near their limit.

With the exception of the community dominated by *Ulex europaeus,* which is local and has not been studied, there are four major easily recognisable plant communities forming a somewhat bewildering patchwork, whose pattern formed the subject of the first ecological investigation of Breckland. Farrow in a series of illuminating papers[1] dismissed soil variability as the primary cause, and from experimental and detailed observational evidence he explained the pattern in terms of the intensity of rabbit-grazing. All the dominants except *Pteridium* are grazed. Their palatability and power of withstanding grazing vary, and the differential effects of diminishing intensity of grazing can be seen in a series of zones with grass-heath the most heavily grazed, followed by a zone of *Carex,* and that in turn by *Calluna.*

In interpreting the vegetation of Breckland, the importance of the biotic factor must be recognised, but too great emphasis upon it obscures primary relationships between the different dominants. By taking cognisance of soil variation and the varying behaviour of the dominants on different soil types, the way is opened to a more exact understanding of plant behaviour and the distribution of the plant communities. The soils of Breckland have this in common that their physical properties vary within a rather narrow range. Open, porous, with a high percentage of coarse particles, and with an almost negligible amount of silt and clay, they have a low water-holding capacity, although this varies with the amount of chalk stones present. And primarily because of the chalk there is considerable chemical variation.

The soil over much of Breckland is derived from the chalky boulder clay, which contains roughly 50 per cent of $CaCO_3$ and 50 per cent of sand with small amounts of silt and clay. By leaching, the $CaCO_3$ is removed from the surface downwards. Following its removal the change in acidity brings about the initiation of podsolisation, the leaching of bases, the mobilisation of the sesquioxides of iron and aluminium and their transference to lower layers. These changes result in a complete series of stages in the development of a podsol, from shallow and highly calcareous

[1] E. P. Farrow, *Plant-Life on East Anglian Heaths* (1925).

soils at one extreme to well-developed podsols at the other. Seven stages in this series may be recognised.

Besides the soil variation brought about in this way, there is a further variation resulting from erosion. The leached soils, having lost their binding material, and supporting a vegetation inadequate to maintain stability, are eroded, often in the form of blow-outs, thereby exposing at the surface different horizons of the podsol profile.[1] The transported sand forms a blanket of variable thickness covering considerable areas and overlying intact as well as truncated profiles.

## GRASSLAND TYPES

The recognition of soil variation throws great light upon the distribution and behaviour of the four major communities. They can be illustrated by a brief account of the variation shown by grass-heath on the seven stages in the development of a podsol, and by reference to the communities dominated by *Calluna*, *Carex*, and *Pteridium*. The "grass-heaths" (grasslands) and the corresponding soils are provisionally designated by the letters A to G: these symbols have nothing to do with the notation used in soil science.

The chief features of these seven stages are summarised on the following page. The perfectness of the series is spoiled in the last four members by the deposition of blown sand, but the soil has been stable for some time and the blanket of sand seems to have assumed properties appropriate to the underlying soil. The first five stages show a well-marked gradient of fertility: F and G are similar to E. The grassland communities described occur on Lakenheath Warren. They are all heavily grazed by rabbits and are thus comparable within themselves; and they differ in some important respects from ungrazed grassland.

*Grassland A.* The highly calcareous shallow soil bears an open vegetation of species tolerant of chalk or exclusive to it. *Festuca ovina* is the most abundant species, *Agrostis* is occasional only. Several species are confined or almost confined to this type: *Botrychium lunaria*, *Calamintha acinos*, *Galium anglicum*, *Ditrichum flexicaule* var. *densum*, *Bilimbia aromatica*, *Lecanora lentigera*, *Placodium fulgens*, and *Psora decipiens*. There are no liverworts. Locally there is more sand, and *Cladonia silvatica* occurs.

*Grassland B.* Of all seven types, this is the richest in species, and its close turf is the nearest approach to chalk pasture found in Breckland. Characteristic species include *Avena pratensis*, *Arabis hirsuta*, *Cirsium acaule*, *Daucus carota*, and *Hypnum chrysophyllum*. The bulk of the turf consists of

[1] A. S. Watt, "Studies in the Ecology of Breckland. II. On the origin and development of blow-outs", *Jour. Ecol.* xxv, 91 (1937).

*The seven stages in the development of a podsol in Breckland.*

| Stage in development of profile | A | B | C | D | E | F | G |
| --- | --- | --- | --- | --- | --- | --- | --- |
| % $CaCO_3$ in surface 6 in. soil sample | 17·90 | 1·610 | 0·129 | 0·00 | 0·00 | 0·00 | 0·00 |
| *p*H in surface 6 in. soil sample | 8·20 | 7·81 | 6·18 | 4·36 | 3·95 | 3·77 | 3·82 |
| % humus in surface 6 in. soil sample (=C×1·724) | 1·551 | 2·129 | 1·962 | 2·198 | 2·482 | 4·074 | 4·046 |
| Exchangeable Ca in M.E. in surface 6 in. soil sample | 51·20 | 34·37 | 6·29 | 0·76 | 0·13 | 0·07 | 0·00 |
| Average depth in inches of soil over chalky boulder clay | 7·5 | 13 | 18 | 34 | Over 60 | Over 60 | Over 60 |

General notes on the profile:

A. Highly calcareous throughout.
B. Calcareous throughout; $CaCO_3$ low at the surface increasing rapidly downwards.
C. Surface soil slightly bleached: a reddish band above $CaCO_3$-containing lower layers.
D. A brown forest soil, overlaid by blown sand slightly podsolised.
E, F, G. Podsol profiles in three stages of development, the last a well-marked podsol: all three overlaid by blown sand.

*Data of the total number of species, of significant species, and of annuals, in seven variants of grassland.*

| Stage | A | B | C | D | E | F | G | A–G |
| --- | --- | --- | --- | --- | --- | --- | --- | --- |
| Total number of species of vascular plants | 50 | 80 | 59 | 37 | 22 | 16 | 9 | 92 |
| Total number of species of bryophytes | 10 | 31 | 32 | 15 | 11 | 12 | 8 | 42 |
| Total number of species of lichens | 15 | 11 | 12 | 12 | 13 | 12 | 12 | 24 |
| Number of significant species of vascular plants | 28 | 49 | 34 | 21 | 8 | 8 | 8 | 55 |
| Number of significant species of bryophytes | 7 | 16 | 11 | 12 | 6 | 5 | 6 | 26 |
| Number of significant species of lichens | 10 | 7 | 8 | 9 | 8 | 7 | 7 | 16 |
| Number of annuals (higher plants) | 20 | 31 | 22 | 13 | 10 | 6 | 2 | 36 |

*Festuca ovina* and *Agrostis* spp.; but an effective, and sometimes colourful, contribution is made by *Asperula cynanchica, Astragalus danicus, Campanula rotundifolia, Carex praecox, C. ericetorum, Galium verum, Koeleria gracilis, Linum catharticum, Lotus corniculatus,* and *Thymus serpyllum*. There are many species of bryophytes but both they and the lichens play a subsidiary part. *Cladonia silvatica* is frequent as "individuals", but never forms pure patches.

*Grassland C.* This type, briefly, is Grassland B without its large calcicolous element; but many exacting species remain. The surface soil is now acid, and *Galium saxatile, Rumex acetosella, Teesdalia nudicaulis* and *Hypnum schreberi* appear in considerable numbers. The turf is more grassy and coarser than in B: *Festuca ovina* and *Agrostis* spp. make up its bulk, but *Campanula rotundifolia, Carex praecox,* and *Galium verum*, are frequent to abundant. There are many species of bryophytes including the characteristic *Bryum roseum*. The most abundant lichen is *Cladonia silvatica*, and it occasionally forms small pure patches under which are found the dead remains of higher plants.

*Grassland D.* A further drop in soil fertility is reflected in the absence of many exacting species, leaving only thirty-seven higher plants. Moreover, the grassy turf is not continuous, and the vegetational cover consists essentially of patches of higher plants and patches of lichen with *Cladonia silvatica* dominant. Again, *Festuca ovina* and *Agrostis* spp. are the chief plants: the relatively exacting *Campanula rotundifolia* and *Galium verum* are less frequent than in C, while the calcifuges *Galium saxatile, Rumex acetosella* and *Teesdalia nudicaulis* are more frequent than in C. There are many fewer species of bryophytes.

*Grasslands E, F and G.* These three types are essentially the same. Eight tolerant significant species of higher plants grow in patches, or scattered in a carpet of lichen composed almost entirely of *Cladonia silvatica*. These species are *Agrostis* spp., *Aira praecox, Festuca ovina, Galium saxatile, Luzula campestris, Rumex acetosella* and *Teesdalia nudicaulis*. The greater number of species (higher plants and bryophytes) which differentiate E and F from G are found largely on soil thrown up by rabbits, thus expressing the finer chemical differences in the three stages of podsolisation represented.

Everywhere, underneath the lichen carpet, occur the remains of grasses, *Rumex, Luzula,* and occasionally *Calluna vulgaris*.

A study of the full lists of species and their distribution among the types of grassland brings out very clearly that from B to G the flora is an attenuating one: there is a fractional elimination of species rather than a radical change. First the calcicoles go; then the more exacting, followed by the less exacting, until finally in G only the tolerants survive. Only a small part of the change is due to the appearance of calcifuges. *Deschampsia flexuosa*, *Nardus stricta* and *Potentilla erecta*, although present in Breckland, are absent from E, F, and G, and also from large areas of Breckland where the soils are certainly acid enough for them. The intimate relation between the soil and the grassland community it bears is thus established. Similar relationships can be seen in Breckland between the soil and communities of bracken, heather, and sand sedge.

## CYCLIC PHENOMENA

*Grass-heath*. In Grassland G, local disintegration of the lichen mat leaves the soil exposed. The rebuilding of the plant cover is initiated by *Aira praecox*, *Festuca ovina* (seedlings) and *Agrostis* spp. (vegetative spread). These afford anchorage for the lichens, which re-establish a continuous cover. In time, the grasses at the centre of a patch die, and death spreads centrifugally until the lichen mat, after a temporary dominance (which may last some years), disrupts.

This cycle of change is a feature of Breckland, and can be well seen in the plant-succession upon bare almost humus-free soil exposed by local erosion in the form of blow-outs.[1] Details cannot be given here, but periodically during the succession there is built up a stage with *Festuca ovina* and *Agrostis* spp. set in a carpet of lichen. As in Grassland G, the grasses die from the centre of a patch outwards, and the pure lichen carpet eventually disintegrates, exposing the soil to erosion. On a partially eroded soil the full succession is telescoped; a new lichen-grass community is built up only to disintegrate once again; and the wave-like advance is repeated until a relatively stable grass-heath emerges upon soil containing about 3 per cent of humus.

But even in a relatively stable grass-heath there is variation from place to place and from year to year. Thus the number of *Agrostis* shoots counted per square metre, in fourteen plots of 0·05 sq.m. selected at random over a uniform grass-heath was 2320, 870, 3004 in the years 1935, 1936, and 1937 respectively: and in two other types of grass-heath the same sequence was obtained. Over the same period the number of shoots of *Rumex* varied inversely, 263, 1299, and 276.

*Calluna vulgaris*, absent from Grassland A, is present on the remaining soil types, and is capable of assuming dominance from C to G. Its ultimate height varies from about 6 in. to about 30 in. according to the soil. The plant lives to an age of about twenty-five years, and (it is important to note) on the poorer soils, at least, the *Calluna* community goes through a

[1] A. S. Watt, "Studies in the Ecology of Breckland. III. The origin and development of the Festuco-Agrostidetum on eroded sand", *Jour. Ecol.* xxvi, 1 (1938).

cycle of change—a stage of invasion, followed by dominance for a number of years, then widespread death. During the tenure of the ground, humus accumulates forming a black peaty *mor* up to 2 in. thick. With the death of the heather and the decay of its stems, the *mor* disintegrates and erosion exposes the mineral soil, which itself may be eroded until the process is checked by the accumulation of flints forming an erosion pavement. On this bare soil a series of communities leads eventually to the establishment of grass-heath, which, if the biotic factor permits, is invaded and replaced by heather. Here there is a retreat of heather for which rabbits are not responsible. Large areas of grass-heath of the poorer types occur, where the only evidence of the former dominance of *Calluna* is the occasional dead stems under the lichen carpet together with the purple stain typical of *Calluna*-heath soils.

*Pteridium aquilinum.* The distribution of bracken in Breckland very strongly suggests a spread from nuclei moist enough for its establishment by spores. Large circumscribed areas of bracken contain either woods, or houses surrounded by trees, from which the spread may have taken place. It is excluded from some areas of cold-air drainage by frost, or it invades them marginally only with extreme slowness, and its vigour varies with the microclimate. But even on soils with similar microclimates variation in height and behaviour is found. Bracken grows in all the seven stages; in height, it varies from approximately 15–18 in. on soil A, increasing through stages B and C to a maximum of about 50 in. in stage D, falling again to about 14 in. in stage G. Incidentally, on the same soil type there is variation from year to year. The bracken also shows a curious patchiness, the patches varying in size from soil type to soil type, but in any one type forming a series in a cycle of change. Some patches have few, deep-set fronds; others have more numerous taller fronds with the part of the petioles showing above ground of intermediate length; while still others carry dense tall fronds with long petioles. In series the average depth of origin of the fronds in the soil becomes less and less. In the last type, death spreads centrifugally from the centre of a patch outwards, and the vacated ground is occupied once more by a scattered population of fronds deep set in the soil.

It has been shown that the rhizome system of bracken is sympodial and that numerous relatively small individual plants make up an area of bracken. It is a typical travelling geophyte: as the rhizome advances in front, it dies away behind, throwing off live branches which thus become independent plants. The number of fronds carried by any one plant is small—approximately one frond to 8 ft. of rhizome.

*Carex arenaria.* The main areas of *Carex* in Breckland lie in the parts of the Little Ouse and Lark Valleys next the Fenland, and upon the blown sand between the large blow-out on Lakenheath Warren and the village of Santon Downham. But it is widely distributed in small and large patches, and it grows, although it does not necessarily become dominant, on all the seven soil stages.

Light has been thrown on this interesting distribution by Mr C. E. M. Tidmarsh,[1] of the Botany School, Cambridge, who has shown that for the successful germination of the seeds a continuous 12 to 20 days' water supply (depending on the temperature) is necessary, and that for the successful establishment of the seedlings similar moist conditions are needed. These requirements limit the establishment of *Carex* to the neighbourhood of water—of rivers like the Lark and Little Ouse, of meres, or of temporary (but not too transient) bodies of water appearing in lower-lying parts when the water table is high.[2] Even if these conditions are satisfied, the establishment of a seedling will be checked by rapid recession of the water table leaving the soil too dry for its survival: thus temporary water-logging offers a somewhat precarious start for *Carex*.

From these *points d'appui*, *Carex* spreads to soils that are essentially dry. The recognition of its early behaviour (the retention by seed and seedling of needs that once may also have characterised the adult), not only explains much of its distribution on Breckland, but also its development on sand dunes near the coast, where it becomes established first in the slacks and later spreads to the dunes. Once established, it spreads freely by rhizomes, and most successfully on loose soil. There are, however, patches of *Carex* in Breckland whose relation to a place suitable for its establishment is not clear. These may be scattered vestiges of a former continuous area in which retrogression has taken place through the activity of rabbits. Just how far *Carex* may degenerate, like *Calluna* and *Pteridium*, without the help of animals like rabbits and mice, is not known. Its behaviour on sand dunes along the coast certainly suggests a loss of vigour with age, but whether this proceeds to the point of annihilation in small or large patches is not yet determined.

The varying height of bracken from year to year; the changing density of *Agrostis*; the results from the application of water to *Agrostis* during dry years; the negative results obtained during the abnormally wet year 1937; and the periodic phenomena already described—all these suggest a causal relation with climate and, in particular, with the rainfall. There is abundant evidence suggesting that scarcity of water is a major difficulty to plant life in Breckland. But the relation between the cyclic phenomena and rainfall

[1] In unpublished work.

[2] As, for example, in the spring of 1937 after a long wet spell.

is not simple, for at any one time cycles in all stages are found, and the period of the rhythm varies from species to species. Up to the present, the data suggest a rhythm explicable partly in terms of the structure and biology of the species, and partly in terms of the effect produced by its own accumulated humus and litter (and for *Agrostis* by the carpet of lichen) on the penetration of rainfall during the summer, when the absolute rainfall is low and the evaporation high. Reversal of the soil-moisture gradient in summer is, in fact, a common occurrence. It is relevant to note that *Agrostis*, *Calluna*, *Pteridium*, and *Carex* form a series with increasing rooting depths; in suitable soils the roots of *Carex* descend to 11 ft., and it is the only species for which a cycle of change has not been demonstrated.

## SUCCESSIONAL RELATIONSHIPS

In the edaphic series A to G, nothing has been said about the causes of the change. This does not imply that long-continued occupation of the ground by grassland could bring about the change from a calcareous soil to a well-developed podsol. On the other hand, the presence of the remains of *Calluna* in D, E, F, and G, and the purple colour of the soils in E, F, and G (slight stain in D), strongly suggest podsolisation under heather, although the heather no longer dominates. In other places, eroded and similarly free from heather, truncated podsols with recognisable remains of heather have been sealed up by a deposit of blown sand. It may be, therefore, that heather was much more widespread than it is now. What is put forward here as a working hypothesis is that leaching has proceeded to produce a brown forest soil, whose further change to a podsol is the work of *Calluna*, that may, in the last analysis, be dominant owing to man.

The varying behaviour of the dominants of the four major communities on the different soils makes it impossible to put forward any simple scheme outlining their relationships to each other. Further work is needed. At the moment, all that can usefully be said is that the relations *Festuca-Agrostis*/*Carex*, *Calluna*, *Pteridium*, and *Carex*/*Calluna*, *Pteridium*, and *Calluna*/*Pteridium*, vary according to the soil, and for bracken, at least, according to microclimate.

But the work of E. P. Farrow[1] and the facts presented in this account make it plain that these communities are not in stable equilibrium with their inorganic environment. The exclusion of rabbits is followed by vegetational change. In different parts of Breckland there are places free, or relatively free, from rabbits and on these areas woody plants have colonised. Of shrubs, the most important is the gorse (*Ulex europaeus*), and of trees the most important are pine, oak and birch. There is little doubt that on certain soils, at least, woodland of some kind would eventually be formed.

[1] E. P. Farrow, *Plant-Life on East Anglian Heaths* (1925).

# INDEX OF PLACE-NAMES

No attempt has been made to include a comprehensive index. Owing to the variety of topics, the list of Contents provides the most convenient reference to subject matter.

## A. CAMBRIDGESHIRE PLACE-NAMES

## B. OTHER PLACE-NAMES

For EU product safety concerns, contact us at Calle de José Abascal, 56–1°, 28003 Madrid, Spain or eugpsr@cambridge.org.

www.ingramcontent.com/pod-product-compliance
Ingram Content Group UK Ltd.
Pitfield, Milton Keynes, MK11 3LW, UK
UKHW010339140625
459647UK00010B/708